THE
LUFTWAFFE

THE LUFTWAFFE

A photographic record
1919–1945

Karl Ries

Translated from the German by
Alex Vanags-Baginskis

AERO
A division of TAB BOOKS Inc.
Blue Ridge Summit, PA

Translated from the book *Luftwaffe Photo-Report* by Dipl. Ing. Karl Ries, published by Motorbuch-Verlag, Stuttgart. Copyright © Motorbuch-Verlag, Stuttgart 1983

This translation first published 1987

ISBN 0-8306-8384-4

Typeset in Monophoto Century Schoolbook by Vision Typesetting, Manchester
Printed in Great Britain by
Anchor Brendon Ltd, Tiptree, Essex

for the publishers
TAB BOOKS, Inc.
13311 Monterey Lane
Blue Ridge Summit, PA 17294 - 0850

LIBRARY OF CONGRESS
Library of Congress Cataloging in Publication Data

Ries, Karl.
   The Luftwaffe : a photographic record, 1919-1945 / by Karl Ries.
      p.   cm.
    ISBN 0–8306–8384–4 : $24.95
      1. Germany. Luftwaffe—History—Pictorial works. 2. Aeronautics.
Military—Germany—History—Pictorial works. 3. Spain. Ejército
Nacional. Legión Cóndor—History—Pictorial works.   I. Title.
UG635.G3R55175   1988
358.4'00943—dc 19   88–6058
                                                                            CIP

Contents

Introduction   7

1. The Luftwaffe as an underground force   9

2. Build-up of a war-establishment Luftwaffe   25

3. The Condor Legion in Spain   43

4. The Polish Campaign   63

5. The 'Phoney War' in the West and Operation 'Weser Exercise'   74

6. The Campaign in France and the aerial war against England   87

7. The Balkans, Greece, Crete and the Mediterranean   117

8. War on the Eastern Front   130

9. North Africa and the German retreat from the Mediterranean   169

10. War in the air over Germany and the occupied Western regions   197

Luftwaffe formations: an explanatory note   230

German abbreviations used in the text   231

Comparative airforce ranks (Luftwaffe/RAF/USAAF/Soviet VVS)   232

Introduction

The action of a highly technical part of the armed forces such as an air arm can be compared to a clock mechanism: with the failure of even the smallest cogwheel one waits in vain for the ticking, the sign that the whole is functioning.

For that reason, this book is not intended as a showcase only for the 'big names'; rather, an attempt has been made also to depict the equally important activities of the lower echelons usually neglected in the literature. The result is a picture closer to reality, free of hackneyed conceptions, and more factual, which should also help to clear up certain glorified notions. The origins, rise, trial by combat and finally decline of this branch of the German armed forces is shown as it happened and not as seen through rose-tinted spectacles.

Even with an extreme effort, the period 1919 to 1939 was an exceedingly short time in which to evolve an air force from practically nothing into a powerful weapon equal to those of other countries. Under the Treaty of Versailles, it was not until 1930 that plans regarding the creation of some flying formations, already prepared by the RWM (*Reichswehrministerium*, State Ministry of Defence) in the mid-1920s, could begin to be realized. This shrinks the 20 years to a lean nine. Considering these facts, combined with the initially quite inadequate budget allocations, it is no wonder that the term '*Risiko-Luftwaffe*' (Risk air force) was used until about 1937.

The appearance of the Luftwaffe, until then built-up secretly – so to speak, 'behind the boarding', before the German and foreign public on 1 March 1935 also brought with it increased exaggeration in the propaganda. With that, the term 'Bluff-Luftwaffe', already used internally, would likewise fit it, although even Western experts have occasionally greatly exaggerated the air rearmament of the Third Reich until the present day.

Be that as it may, the enormous financial demands of the Luftwaffe and its ancillary industry, which almost broke the State budget, finally resulted in an air force that was more than equal to those of the potential enemy countries before the war. Whatever had been right or wrong in planning until that time would be revealed during the coming six years of war. Where the responsible powers-that-be were certainly wrong was their estimate of the enemy's industrial capacity, something the German leadership clung to right into the third year of the war. The mistaken notion of resisting the inexhaustible reservoir of raw materials on the opposing side with isolated outstanding German achievements in aviation and rocket technology held sway until the final phase of the Second World War. It seems likely the German leadership was as aware of this false optimism as were the various branches of the RLM (*Reichsluftfahrtministerium*, State Ministry of Aviation) with their research-, development- and procurement-offices which, according to the minutes of countless meetings, expressed doubts about the so-called '*Sofort-*' and '*Notprogramme*' (Immediate and Exigency programme) plans.

Closely interwoven with the spheres of armament and technology, the Luftwaffe had to face up to – or even take initiatives against – a materially superior enemy with the equipment it had available. The tragedy of this almost impossible task, to assert itsef on many battlefields against superior powers, is evidenced by the personnel and material losses of the Luftwaffe. In this respect, the war diaries, operational and combat reports speak a cold and sober language compared to the 'Jingo'-patriotism of the contemporary propaganda machine. Whoever today still sees and judges the air war through the optics of the Luftwaffe periodical *Der Adler* has to learn to rethink. Such products of the official press were more or less factual until about mid-1940 but afterwards were not worth the paper they were printed on.

This book is an attempt to present the happenings before 1945 in pictures that reflect the real life of the Luftwaffe. I have endeavoured to bring to the reader that characteristic 87-octane airfield atmosphere which all those who participated at that time are sure still to feel in their nostrils.

I am greatly indebted to my wide circle of acquaintances among the former members of the Luftwaffe, from the Weapons General to the ordinary 'cravate soldier'*. After all, it was conversations with these former 'actors' in that drama that provided the foundations on which such a condensed representation in pictures could be set up in the first place. To mention even one name here would be unfair; to mention them all would be impossible. For that reason, I make an all-round bow to all those who had worn the Luftwaffe 'blue' and did not slam the door in my face.

I am also obliged to the *Bundesarchiv/Militärarchiv* at Freiburg and the *Deutsche Dienstelle WAST* [German Armed Forces Information Office] in Berlin for their unbureaucratic collaboration and help.

Mainz-Finthen
KARL RIES

*Disparaging term for the Luftwaffe soldiers used by the Army in particular due to the fact that even 'other ranks' of the Luftwaffe could wear a tie with their going-out uniform. Occasionally (like here) also used by the Luftwaffe personnel as a matter of pride and distinction.

1. The Luftwaffe as an underground force

With the end of the First World War – strictly speaking with the signing of the Versailles Treaty on 28 June 1919 – the Imperial German Air Force ceased to exist.

Although the Air Force leadership tried to move the squadrons and wings back into Germany in an orderly and disciplined way, various revolutionary activities resulted in some aircraft, aero engines and maintenance tools getting into the hands of private purchasers and, in some cases, foreign customers, soon after the Armistice. Under such circumstances the demand of the victorious powers that Germany should deliver 2600 military aircraft and the same number of aero engines could not be fulfilled, and an agreement was finally reached stipulating the delivery of 1700 fighters, bombers and reconnaisance aircraft and 2000 aero engines.

As became known later, quite a few military aircraft were clandestinely 'redirected' from collecting centres to remote sheds and barns to preserve them for future German aviation efforts. However, this pious hope was frustrated by the formation of the Interallied Aviation Control Commission whose task was to supervise the strict observance and implementation of Versailles Treaty articles relating to aviation. In addition to Germany being debarred from having an air force, these injunctions also imposed strict limitations on the engine power, speed and all-up weight of civil aircraft being constructed in Germany. Apart from that, no aircraft was allowed to fly without permission granted by the IAAC.

Naturally, all this amounted to a challenge to evade the regulations. If it did not want to change its working structure fundamentally, the German aviation industry just had to find another way out.

The opportunity came with the lifting of the general prohibition of aircraft construction in Germany in May 1922, which applied to all old and new aircraft manufacturing facilities, but with the proviso that only light aircraft could be built in Germany. If there was to be no lagging behind foreign competition as regards development and manufacturing technology, the only way of evolving aircraft for which there was a worldwide demand (commercial aircraft, transports and seaplanes) was to develop and build them abroad. As a result, in the early 1920s several German aircraft manufacturers established branches in other countries, such as Junkers at Limhamn in Sweden, Fili in the Soviet Union, Angora in Turkey and Rohrbach in Denmark, while Dornier found new facilities at Altenrhein in Switzerland, Pisa in Italy, Kobe in Japan and Papendrecht in Holland. The Heinkel works remained at Travemünde, but took on orders from the USA, Japan and Sweden. Thanks to this private initiative several efficient aircraft manufacturing facilities familiar with the latest technological developments remained in being in Germany.

After the attempts by the *Reichswehrministerium* (RWM, State Defence Ministry) to maintain some small flying units described as 'police air squadrons' into the postwar years had failed, a new office camouflaged as *Luftschutzreferat* (Air Defence Department) was founded under *Hptm* Wilberg in March 1920 with the brief of planning a modest air force and its equipment. The first task of this new department in the RWM was to save as many existing airfields from demolition as possible so that they would be at least available for flying training. However, it was not until spring 1923 that the Sportflug GmbH (Sports Flying Ltd) could be founded. Supported by the RWM, this new organization was intended to provide refresher and *ab initio* training for pilots on a total of ten airfields. These training courses began that same spring.

Another camouflaged training location appeared in summer 1926 when the Deutsche Luftfahrt GmbH (German Aviation Ltd) began to function from some sheds at Böblingen and Würzburg. From late 1933 on, these two firms evolved into the so-called *Fliegerübungsstellen* (Pilot Practice Establishments). Until that time, Wilberg's LSR had devoted itself principally to the task of planning an air force, keeping in touch with the aviation factories, and a more intensified furthering of the gliding movement in general and the Rhön competitions in particular.

In 1924, the *Reichsmarine* (German Navy) established a clandestine naval aircrew training school at Kiel-Holtenau. Known as the Severa GmbH, this organization was less involved in actual training than in flying target simulation for the anti-aircraft defences of the German warships. On the other hand, at the Severa GmbH branch at the Norderney seaplane station flying training had absolute priority.

While flying schools operated by the Sportflug GmbH and Deutsche Luftfahrt GmbH could only train the pilots to A2 grade, the Deutsche Verkehrsfliegerschule (DVS, German Commercial Flying School) offered them an opportunity of progressive training on larger aircraft to qualify for the B and C categories.*

The DVS, originally based only at Berlin-Staaken, later had branches at Schleissheim, Braunschweig (Brunswick), Warnemünde and List on Sylt island. It had been founded in 1925 on the pretext of training pilots for service with the various commercial flying enterprises. In actual fact, even from the early years its activities could be described as those of a paramilitary flying training centre. Where else but in an air force would all those hundreds of pilots who had successfully passed the DVS training have found employment?

On the other hand, overt military training of pilots was impossible within Germany, due to the constant surveillance of the victorious powers. Of necessity, such training had to be carried out abroad, in a country over which the Allies had no influence or right of inspection – the Soviet Union.

The starting point here was the Treaty of Rapallo, signed between the Soviet Union and Germany on 16 April 1922, which created the opportunity for discussions along these lines. The agreement regarding the use of the Soviet airfield at Lipetsk, north of Voronezh, was finalized by the *Reichswehr* and the Red Army in April 1925. From June of that year, this airfield and part of its installations were at *Reichswehr* disposal for the training of fighter pilots and observers as well as the testing of military aircraft of German construction de-

veloped and built outside Germany. The flying training school, known as *Fliegerschule Stahr* (after its first commanding officer, *Major* Stahr) began its activities in summer 1925 after the delivery of 50 Fokker D XIII fighters via a bogus firm.

From the very beginning, the *Reichswehr* – in this case the *Reichsheer* (Army) – was endeavouring not to let any NCOs or officers from its limited cadres (in strict adherence to the Versailles Treaty) join the new arm, then still non-existent, urging instead the recruitment of flying personnel from civilians and reservists.

The Stahr Flying School, later renamed *Flugzentrum Lipetzk* (Lipetsk Flying Centre), was in existence until October 1933. The reasons for its disbandment were in part the low serviceability of the aircraft at Lipetsk and therefore the need for new investment, but also to a considerable degree the change of power in Germany. In late autumn 1933 another two courses of German figher pilots completed their training on Fiat biplanes at Grottaglie in Italy before the fighter training school at Schleissheim became a State-funded establishment. With that, it was possible to start the military training of pilots in Germany proper.

The creation of the actual air force was another matter. An early organizational plan prepared in May 1928 envisaged the following flying units by 1 April 1931:

8 reconnaissance *Staffeln*	@ 6 aircraft	= 48 aircraft
3 bomber *Staffeln*	@ 6 aircraft	= 18 aircraft
4 fighter *Staffeln*	@ 9 aircraft	= 36 aircraft
15 *Staffeln* with		102 aircraft

As predecessors of the later *Jagdgruppen* we must consider the three so-called *Reklamestaffeln* (Advertising Squadrons) based at Berlin-Staaken, Königsberg and Nürnberg-Fürth, as from late 1930 on playing the role of the 'flying arm' of the German armed forces during manoeuvres, although ostensibly they were engaged in dropping advertising leaflets, and were available for hire to private firms as 'sky writers' and for the towing of advertising slogans.

The organizational periods of 1932, 1933 and 1934 set up the framework for the future establishment of the flying troops on a considerably more generous scale, listed the airfields to be completed by the end of each term and at the same time prepared realistic aircraft distribution plans. As a result, when the clandestine Luftwaffe was officially revealed on 1 March 1935, the following flying units

---

*A2, Advanced single-engined, up to 3 seats; B, Single-engined, up to 6 seats; C, Multi-engined.

were already in existence or in the process of formation on the establishment:

6 short-range reconnaissance *Staffeln*
5 long-range reconnaissance *Staffeln*
4 land-based fighter *Staffeln*
1 naval fighter *Staffel*
13 bomber *Staffeln*
1 dive-bomber *Staffel*
and 4 coastal aviation *Staffeln*.

Except for the bomber units, most of the formations had a certain amount of modern equipment. New developments had been commissioned since 1933, and some were already undergoing tests.

Naturally, the controls exercised by the victorious powers of the First World War, even if relaxed, demanded the strictest secrecy regarding these activities, which ran counter to the Versailles Treaty; hence the use of 'harmless' designations for all such 'underground' formations. Who could imagine that the *Forst- und landwirtschaftliches Flugversuchs-Institut* (Forest and Agricultural Experimental Flying Institute) really stood for the Reconnaissance *Staffel* Prenzlau, or the *Hanseatische Fliegerschule* (Hansa Flying School) was really the Bomber *Gruppe* Fassberg? Other examples of such 'cover designations' were the *Reklamestaffel Mitteldeutschland, Funkpeilversuchsinstitut des Reichsverbandes der elektrotechnischen Industrie* (Experimental Radio Direction-finding Institute of the State Association of the Electrical Industry) and *Luftbildlandesvermessung Westdeutschland* (Aerial Photographic Surveying of West Germany), to mention only a few. Until November 1933 this situation was helped by the ruling that all military aviation personnel under training had to wear civilian clothing and outwardly create an impression that they were engaged in civil flying activities. The DLV (German Flying Sport Association) uniforms introduced afterwards remained the apparel of the members of this organization as well as clandestine military aviation personnel until 1 March 1935, when the State Minister of Aviation issued preliminary dress regulations for Luftwaffe personnel (later known as L.Dv.422).

The scene in aircraft factories and at aircraft collecting centres after the end of the First World War: scores of airframes and engines awaiting scrapping. At the DFW (Deutsche Flugzeugwerke) plant in Leipzig-Lindenthal aircraft fuselages were left to rot in the open.

Using rebuilt military aircraft an attempt was made to operate a makeshift air-transport service within Germany. This hurriedly modified Junkers J 10 all-metal close support aircraft was used early in 1919 for shuttle service between Berlin and Weimar, the seat of the National Assembly.

A former Rumpler C I reconnaissance biplane sporting the civil registration D–97 in service with the DVL (Deutsche Versuchsanstalt für Luftfahrt, German Aviation Experimental Establishment) in Berlin-Adlershof.

12

Due to the prohibition of aircraft construction in Germany, German aircraft companies developed their larger aircraft abroad. This Junkers K–37 bomber was built by Flygindutri in Limhamn, Sweden, in 1927.

Dr Claudius Dornier with his design office moved to Pisa in Italy where he developed the basic model of the later famous Wal flying boats. Leading a section of three Military Wals, the Spanish Major Llorente made headlines flying from Melilla in Spain to Santa Isabel in Spanish Guinea.

Temporarily lacking any demand in his own land, Dornier delivered a small series of his three-engined Do Y bombers to Yugoslavia. These aircraft were built in 1931 at Altenrhein in Switzerland.

13

Top Left: *In 1927 Theo Osterkamp, a former Reserve* Leutnant *and a naval fighter ace of the First World War became Station Leader of Severa Flying Enterprises at Kiel-Holtenau, a paramilitary operation that managed to survive into the 1920s. The picture shows Osterkamp (right) with the Secretary of State for Air, Erhard Milch, before the Circuit of Europe Flight in 1934.*

Top right: *Among its other activities the Severa GmbH flew target simulation for the Reichsmarine (German Navy). This is a Severa Junkers W 33be, D–1384, shortly before releasing its towed target sleeve.*

In 1928 the Arado-Flugzeugwerke began production of the W II floatplane to train naval airmen. This type was delivered to the DVS (Deutsche Verkehrsfliegerschule, German Commercial Flying School) at Warnemünde.

Top: *Basic flying training of 'civilian' pilots was carried out at the Deutsch Luftfahrt GmbH (German Aviation Co. Ltd). The above photo shows the hangar of its branch at Böblingen in the late 1920s. Note the warning on the hangar doors. Left: 'No entry allowed without permission!' Right: 'Even with permission, entry at own risk!'*

Below: *The DVL (German Experimental Aviation Establishment) in Berlin-Adlershof was an important scientific institution during the years of forbidden military aviation. This aerial photo shows the layout of the DVL buildings at the south-eastern edge of the Johannisthal airfield.*

Top left: *The training personnel of the Böblingen flying school in 1929. From left: aerobatic instructor Walter Spengler, the school principal Gustav Engwer and flying instructor Hermann Weller. Both Spengler and Weller were killed in a mid-air collision on 28 September 1930.*

Top right: *At the end of 1933 Dipl. Ing. Hermann Huppenbauer, was appointed principal of the flying training centre at Böblingen – which shortly afterwards became a Pilot Practice Establishment (Fliegerausbildungsstelle), training establishment for military flying.*

Bottom: *This Klemm L 20 trainer, stuck in the telephone wires while coming in to land, displays the markings typical of the Deutsche Luftfahrt GmbH aircraft park – a black/white striped tailplane assembly.*

Top: *Powered by an 600 hp BMW VI inline engine the Dornier Merkur was classed as a Category C 1 aircraft and served to familiarize young pilots with heavier machines. Note the reverse registration on the starboard wing.*

Bottom left: *Dr Günther Ziegler was principal of the land branch of the DVS, the German Commercial Flying School. After completion of their basic training the budding pilots learned to fly heavier aircraft at this establishment.*

Bottom right: *Trainee pilot Oskar Henrici posing in front of an Udet U–12b Flamingo at DVS Schleissheim. As a fighter pilot Henrici was killed in Spain on 13 November 1936.*

Top: *Two of the early
flying instructors at
DVS Schleissheim
were the much-
decorated First World
War pilot Walter
Fruhner and his
brother Otto. Whenever
they got involved in an
argument the brothers
would inevitably end
up calling each other
'Knallkop' ('Cracker
head') and 'Arschloch'
('Arsehole'), and were
later known under
these 'pet names'
throughout the
Luftwaffe. Walter is
shown here.*

Centre: *The famous
German aerobatic
champion Willy Stör
was also instructing at
Schleissheim, where he
taught his 'flock' the
gentle art of rolls and
loops. He is seen here
resting against a Raab-
Katzenstein K1 Ic
with a yellow-blue
striped tailplane, the
markings of the DVS.*

Bottom: *The 1932/33
'Jungmärker' ('New
Blood') fellowship of
the Schleissheim flying
training school.
Among these eager
youngsters one can
recognize some later
well-known pilots, such
as von Bothmer,
d'Elsa, Harder,
Kroeck, von Dohna
and Pitcairn.*

18

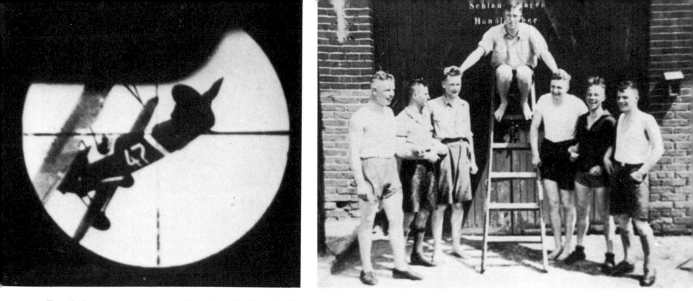

Top left: *A gun camera shot of a Fokker D XIII at Lipetsk.* Top right: *Some of the trainees of the last fighter pilot course (1933) in the Soviet Union. From left: Berlin, Schlichting, Robitzsch, Steen, d'Elsa, Pitcairn and Henrici.*

Centre: *Dutch Fokker D XIII fighters for the Stahr clandestine German flying training school at Lipetsk arrived there by ship from Stettin in May 1926. Shown here is a Fokker D XIII on skis in winter 1927/28.*

Bottom: *The hangar complex of the clandestine German fighter pilot school at Lipetsk, north of Voronezh. Other activities there were the training of observers and flight-testing of covertly built German military aircraft.*

Top: *Encouraged by the German government, Prof Hugo Junkers started a branch factory at Fili, near Moscow, in 1923. One of the products of this new plant was the R–02 (J 20) two-seat reconnaissance monoplane, a military version of the Junkers A 20. Note the Soviet markings.*

Centre: *Another Junkers product at Fili was the K–30 three-engined bomber, shown here in Soviet Air Force markings. It could be fitted with wheel, ski or float landing gear.*

Bottom: *The excellent relations between Germany and the Soviet Union in the 1920s were reflected in numerous contacts with Soviet aviation. On 31 August 1926, a Tupolev ANT–3 landed at Berlin-Tempelhof on an official friendship visit.*

Top: *The Heinkel works at Warnemünde kept in step with foreign developments in the construction of seaplanes and catapults. Heinkel HD 30b floatplane D–1463 is seen undergoing test launches from an experimental floating catapult in 1929.*

Centre: *With the giant 12-engined Do X flying boat, Dornier hoped to interest the German Navy, but no orders materialized. Because of Allied restrictions, this colossus had to be developed and built at Altenrhein in Switzerland.*

Bottom: *Christening ceremony of the large Junkers G 38 passenger aircraft at Berlin-Tempelhof on 29 April 1933. One of two examples in Lufthansa service, D–2500 was requisitioned at the beginning of the Second World War and allocated the military registration GF + GG.*

Top: *During the crisis years the Albatros works kept its head above water with a varied construction programme. Four different types of aircraft are seen in this hangar, the L 101 (D–1895), L 100 (D–1906), L 79 (D–1871) and (without markings) L 83 transport. In 1932 the Albatros works were taken over by Focke-Wulf Flugzeugbau.*

Bottom left: *The Minister of State for Aviation, Hermann Göring, photographed during his visit to an aviation day at Oberwiesenfeld in the summer of 1933. From left: the aerobatic champions Gerd Achgelis and Willy Stör, Göring and Theo Croneis, later Vice-Chairman of Messerschmitt. Such 'air days' were very popular and were organized under the motto 'The German people must become a nation of airmen!'*

Bottom right: *The outstanding aerobatic pilot Robert 'Bobby' Weichel began his career as an instructor at the various training centres which later became training regiments of the Luftwaffe. He is shown here in DLV (German Air Sport Association) uniform as Flight Commander standing next to a Bücker Bü 131 Jungmann trainer.*

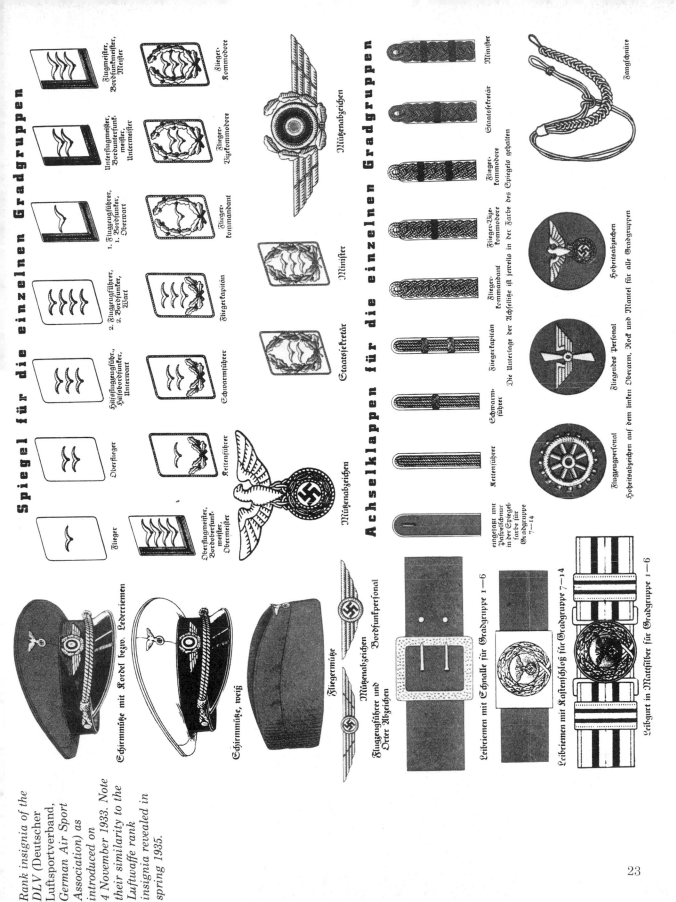

Rank insignia of the DLV (Deutscher Luftsportverband, German Air Sport Association) as introduced on 4 November 1933. Note their similarity to the Luftwaffe rank insignia revealed in spring 1935.

23

Top: *The Heinkel HD 38 was intended to fulfil the fighter role in the clandestine Luftwaffe. It could be fitted with wheel and float landing gear, but it was overtaken by more modern projects and was not put into series production.*

Centre: *On the other hand the Arado works secured a large order from the RWM (Reichswehrministerium, State Ministry of Defence) for their Ar 64D single-seater fighter. These agile biplanes were in service as trainers with the* Fliegergruppe (S) *at Schleissheim, a clandestine fighter pilot training school, until 1935.*

Bottom: *Until its re-equipment with the He 46 late in 1934 the* Fliegerstaffel Cottbus *operated the Albatros L 78 reconnaissance biplane. Before the official 'unveiling' of the Luftwaffe this reconnaissance unit was known as DVS GmbH, Zweigstelle Cottbus (DVS Cottbus Branch).*

2. Build-up of a war-establishment Luftwaffe

The new branch of the German armed forces, the Luftwaffe, was introduced to the German and world press early in March 1935. 'Camouflage' regulations had been revoked on 1 March 1935 and, after the re-introduction of universal military service on 16 March, the way was open for overt rearmament of all three branches of the armed services.

In the case of the Luftwaffe, in addition to a considerable increase in the number of active formations on the 'cell splitting' principle, there was a similar increase in the number of pilot, air armament and blind-flying training schools. In addition, the training and preparation of a Luftwaffe replacement reserve at the so-called *Fliegerübungsstellen des DLV* (DLV Aircrew Practice Centres) which were distributed all over Germany took on a special signficance. In these, DLV members fit for active service, who still wore the DLV uniform (i.e. were not soldiers), were trained as pilots, observers, radio operators and signals personnel to be ready in case of mobilization. In all *Fliegerersatzabteilungen* (FEA, Aircrew Replacement Detachments), and Luftwaffe pilot training and air armament schools, volunteers and others liable for military service had to wear Luftwaffe uniform as a matter of duty.

On 1 October 1936 the number and composition of these training establishments was as follows:

DLV aircrew training and instructional centres	51
Luftwaffe aircrew replacement detachments	14
Luftwaffe pilot flying training schools (land)	9
Luftwaffe pilot flying training schools (sea)	4
Luftwaffe blind flying schools	2
Luftwaffe navigational training course	1
Air armament schools:	
(a) Bombers	3
(b) Reconnaissance	2
(c) Fighters	1
(d) Air armament (sea)	2

The organization of the Luftwaffe flying formations since its 'decamouflage' changed insofar as all *Staffeln*, be they reconnaissance, bomber-, fighter-, dive bomber- or naval-aviation now had a uniform strength of nine aircraft, plus another three listed in reserve.

At the same time the reconnaissance- and naval aviation units were limited to smaller formations of *Gruppe* strength, while all others were intended to be concentrated into *Geschwader* (of three *Gruppen* each), their establishment returns showing the following composition and equipment:

	Aircraft	+ Reserve	= Total
Geschwaderstab with *Stabskette* (HQ Flight)	3	1	4 = 4
I *Gruppe*: *Stabskette*	3	1	4
1. *Staffel*	9	3	12
2. *Staffel*	9	3	12
3. *Staffel*	9	3	12 = 40 aircraft
II *Gruppe* (4., 5. and 6. *Staffeln*)	30	10	= 40 aircraft
III *Gruppe* (7., 8. and 9. *Staffeln*)	30	10	= 40 aircraft
			124 aircraft

Naturally, such formations could not be created out of thin air in such a short time, without the necessary service airfields, aircraft parks, aircraft ordnance departments, ammunition and fuel stores with all their many subordinate depots and service stations. During the years after 1935 Germany was like an enormous construction site with the most modern service airfields and bases literally hatching out all over the place, carefully planned to blend with the landscape. Existing airfield installations were expanded according to modern principles and adapted to the most pressing demands.

If in autumn 1935 there were 57 land and 13 naval air station commands (service air base headquarters), a plan dated 29 August 1935 envisaged an increase to more than double that number by 1 October 1936. In consequence it also became urgently necessary to increase considerably the personnel of the existing six *Luftkreiskommandos* (Air District Commands) as regional headquarters and administration departments.

Technical air rearmament had a similar impetus. Carrying out the many project and development tasks requested by the Technical Office of the RLM (*Reichsluftfahrtministerium*, State Air Ministry) required extensive construction of new aircraft plants, such as the Heinkel works at Rostock-Marienehe and Oranienburg, Dornier at Wismar and Oberpfaffenhofen, Junkers at Bernburg and Aschersleben and many others. Before Ernst Udet became chief of the Technical Office (C-Amt) at the RLM in June 1936, that all-important control and order-placing post had been looked after by *Oberst* Wimmer. The forced schedules imposed on this office by the State led to the compression of the hitherto accepted time span before a new aircraft type was ready for series production, from three to just two years – a fateful decision, as was soon to be evident.

Under such conditions it was inevitable that a technically complicated device such as an aircraft would generally be put into large-scale series production before it was really ready – which, in turn, led to constant modifications and improvements during the production run. This problem was to prove especially grave during the war years, the most notable examples being such aircraft types as the Me 210/410, He 177 and Ju 188; there were others.

Test establishments for all new aircraft types were the former RDLI (*Reichsverband der Deutschen Luftfahrt-Industrie*, National Federation of the German Aviation Industry) centres at Rechlin and Travemünde.

Directly answerable to the Minister of State for Aviation and C-in-C of the Luftwaffe, *General der Flieger* Hermann Göring, were the Inspectors of the various arms of the service (reconnaissance, bombers, fighters, naval aviation, pilot training schools, anti-aircraft artillery and signals), all of them striving to keep pace with the formation plans, but failing time and again in the face of unforeseen problems. Under these circumstances it was extremely risky, after the Allied troops had vacated the Rhineland on 30 June 1930, to order the German Army to march into this officially demilitarized zone 'under cover of the Luftwaffe' on 7 March 1936. Even more astonishing was the fact that other treaty signatories, France, Great Britain and the USA, showed no intention of taking any kind of counter-

measures. The few flying formations – after making wide turns – repeatedly overflew the Rhine, always at different places, supposedly to create the impression of a powerful Luftwaffe. In actual fact, this 'aerial might' consisted of just six *Staffeln*! But the bluff succeeded and, for the first time, made the German people look in admiration at their 'Führer'. Of course, all this also strengthened the Führer's assumption that provided one was prepared to take risks, with a suitable dash of impudence and self-confidence, there was very little chance of serious intervention from outside.

Despite this demonstration of German 'air power' for external effect, the setting-up of new formations lagged constantly behind schedule, as regards both personnel and aircraft equipment, until the beginning of the war in 1939. Likewise, the German building economy had insufficient capacity to master the numerous construction projects started all over the place.

As a result, many of the newly established Luftwaffe *Gruppen* could not carry out any flying practice for months on end because their intended service airfields were still under construction – and they could not move to another airfield because all of them were already 'over-occupied'.

The qualitative and also quantitative defects became especially noticeable at that moment when, following a military *putsch* in Spain in July 1936, General Franco asked the German government for material assistance.

Only the fact that the Luftwaffe fighter and bomber *Gruppen* were about to re-equip with more modern types of aircraft allowed the delivery of superseded older machines, such as the He 51 biplane fighters and Ju 52/3m makeshift bombers, to Spain in 1937. To be sure, later obligations in connection with the formation of the 'Condor Legion', which grew to the size of a *Fliegerkorps*, subsequently demanded the export of more modern aircraft types as well.

From 1935 onwards, in addition to the completed service airfields, various Air District commands also began to concentrate on the construction of Grade I and II operational bases (*Einsatzhafen*, or *E-Hafen*) for deployment in case of mobilization. These operational bases were mostly situated near the borders and intended for the temporary accommodation of Luftwaffe formations. As such, they were only given access to the main communication roads for supply purposes, foregoing any rail connections and even hardened take-off and landing runways. In most cases such operational bases remained in use as ordinary meadows and hayfields until an emergency situation (*Fall A*, or 'Case A') and were used only occasionally as temporary field bases by Luftwaffe formations undergoing mobilization deployment exercises.

Until '*Fall Otto*' ('Case Otto'), the entry of German troops into Austria, there were quite a few such operational bases in being in the southern regions of Germany, used for deployment of various Luftwaffe formations during the preparatory period. Although no serious resistance was ever expected, the Luftwaffe force concentrated before the entry into Austria, which began on 12 March 1938, was quite considerable. The units participating in 'Operation Otto' came from the *Luftwaffengruppe 3* (*General der Flieger* Sperrle and comprised two long-range reconnaissance *Staffeln*, three short-range reconnaissance *Staffeln*, two Stuka-*Staffeln*, one fighter *Gruppe* and one bomber *Gruppe*, as well as parts of Flakregiments 5 and 28. In those weeks leaflet-dropping and demonstration flights were among the orders of the day of most of these Luftwaffe formations. There was also a *Transportgruppe Chamier* equipped with Ju 52/3m transport aircraft which carried troops and material supplies to the airfields around Vienna.

The incorporation of the Austrian Air Force formations and their ground organization into the Luftwaffe not only added well-trained flying personnel but also achieved a favourable strategic situation on the southern flank of Czechoslovakia. In the reviews of the (potential) enemy situation carried out since the end of the First World War the Czech territory had always seemed a thorn in the flank of the German State, in 1936 even gaining the description of 'a forward aircraft carrier of the East'. [Not only was there close collaboration between the Czech and Soviet military at the time but also Czechoslovakia was the only foreign country entrusted with the licenced production of the latest Soviet frontal bomber, the then very advanced Tupolev SB–2, just joining the Soviet Air Force! After the receipt of 30 sample aircraft their production was initiated in Czechoslovakia (with Soviet technical assistance) in 1937, being known as B.71 in Czech service. A total of 111 B.71s were built – including some completed *after* the German occupation. Tr.]

The actual incorporation of the few Austrian aviation companies into the Luftwaffe took place under the control of *Generalmajor* Alexander Löhr, the former commander of the Austrian Air Force. This was followed by an expansion of existing military airfields and the construction of new ones in 'Ostmark', as Austria had been immediately renamed. The

Luftwaffen-Kommando Österreich, later renamed *Luftwaffen-Kdo 4* and based in Vienna, was in charge of the winding up and takeover of Austrian Air Force personnel and material.

In the same year the Luftwaffe had to appear once more as the 'threatening finger', this time during the acquisition of the Czech border areas ceded to Germany in the Munich Agreement. However, in March 1938 the German military leadership had already prepared a study named '*Fall Grün*' ('Case Green'), the occupation of the Sudetenland, which covered the eventuality if there was no peaceful agreement. This study also incorporated '*Fall Rot*' which assumed an intervention by France and England. The continuous extension of the Czech border fortifications, noted since autumn 1937, together with the possible intervention of the other two partners in this agreement, demanded the preparation of considerably stronger troop contingents by both the Army and the Luftwaffe.

Despite the Munich Agreement signed by Chamberlain, Daladier and Hitler on 29 September 1938, German military measures remained unchanged. The password '*Geschwadertag Albert*' ordering earmarked Luftwaffe units to occupy their operational bases had been given the day before the agreement was signed. In other words, German troops would certainly have been ordered to move into Sudentenland even without the Munich agreement – and even if it meant the danger of a worldwide conflict.

The main body of the Luftwaffe formations was provided by the *Luftwaffengruppenkommandos* 1, 3 and '*Osterreich*', with a total operational strength of 1300–1400 aircraft including reserves, the main effort being concentrated in Upper Silesia, Northern Bavaria and the Austrian border areas. However, the operational activities of the Luftwaffe were limited to the exploration of airfields, keeping an eye on the advancing German troops, propaganda flights and transfer to airfields in the occupied area. The occupation of Sudetenland was completed according to schedule by 10 October 1938, with no need for the Luftwaffe to take a hand.

After an international arbitration had requested that other areas of the remaining Czech state should be allocated to Poland and Hungary, the Slovaks formed an autonomous government in their region which almost immediately led to renewed trouble between the Czech and Slovak populations. Due to excesses against the ethnic Germans this situation gave the German government a pretext to put the remaining Czech territory 'under German protection' – particularly as Hitler could also refer to pleas for assistance from the Slovak president Dr Tiso.

And so it happened that on 15 March 1939, only a few months after the occupation of Sudetenland, German troops crossed the border into the Czech heartland and Luftwaffe formations approached the well-laid-out and still partly snow-covered Czech airfields around Prague, Prossnitz and Olmütz (Olomovc). But, this second part of 'Operation Green also did not require any action by the Luftwaffe: the well-trained and well-equipped Czech Air Force units, so highly rated by military experts, offered no resistance. About 1000 aircraft of various types fell into the German hands, some to be passed on to the Slovak squadrons, others to be transferred to the various flying training schools in Germany. [Most of the modern B.71/Tupolev SB–2 bombers were impressed into Luftwaffe service as target tugs, and some as bomber trainers; others were passed on to the Bulgarian Air Force. By then, the type had already gained fame in Spain where, flown by Soviet aircrews, it became known as 'Katiuska'. *Tr.*]

This breach of the Munich Agreement was followed just a week later, on 23 March 1939, by the entry of German troops into the Memel (Klaipeda) district in Lithuania. This action was justified by the German government under international law as a long-overdue reunification of the Memel strip with the old Motherland.

These repeated occupations of foreign territories naturally led to increased anxiety about the threat of war in Europe. In the same month, Britain introduced universal military service and, together with France, gave Poland a firm guarantee, followed by similar guarantees to Romania and Greece in April 1939.

That year the annual autumn manoeuvres of the German armed forces, including plans for strong Luftwaffe participation, fell into abeyance. Apart from a few war games under the covernames '*Generalstabsreise 1939*' and '*Donau*' (General Staff Journey 1939 and Danube) the principal attention of the Operations Staff of the German armed forces remained directed towards a speedy establishment of combat readiness in case of mobilization. This was especially so of the Luftwaffe whose voluntary 'Condor Legion', after the end of the Spanish Civil War on 28 March 1939, had returned to Germany in May. Immediately afterwards, on RLM orders, all formation leaders had to prepare reports outlining tactical and technical experiences in Spain. These then had to be incorporated into the official Luftwaffe service regulations (LDv) and issued to the troops as quickly as possible.

During those months of increasing diplomatic activity all over Europe the Luftwaffe command was busy on preliminary work at the various operational bases along the eastern and western frontiers of Germany. Detachments of the State Labour Service (RAD) worked around the clock to tidy up matters of detail. Some of this involved the removal and assembly of the accommodation barrack components stacked in storehouses camouflaged as barns alongside the operational bases, and the fitting of operational command facilities in the local farmer's house. Special units were busy completing the airfield lighting installations, while advanced detachments, quartermaster troops and fuel transport columns were assembled in the central airfields. Simultaneously, the empty anti-aircraft positions around the operational bases were filled by gunners and their weapons.

Growing political tension, harsh words and the bogged-down situation of the nations of Europe could signal only one thing: war!

Top left: *Walter Rubensdörffer, one of the instructors at the Schleissheim fighter training school, demonstrating his skill in closest formation flying with a Fw 56 Stösser. The military unit markings were openly displayed after the covert regulations were dropped on 1 March 1935.*

Top right: *A flight of He 51A–0 pre-production series of fighters. Until the official introduction of military markings in May 1936, the fighter squadrons flew aircraft with civil registrations.*

Bottom: *With the arrival of the He 51, the Arado Ar 65 interim fighters were relegated to fighter training schools. This manoeuvrable but rather slow biplane could be found flying in that role right into the war years.*

Top: *By purchasing two Curtiss H–81 Hawk single-seaters of proven diving ability Göring intended to persuade Ernst Udet, an enthusiastic aerobatic pilot, to take over an important post in the newly-created RLM (State Air Ministry). Both aircraft were put at Udet's disposal who is seen here taxing the D–3165 before taking off in Berlin in 1934.* [*The aircraft were actually Curtiss F11C–2 Hawk II Goshawk 'Helldiver'. After witnessing an impressive demonstration of their 'dive bombing' with sand bags in the USA in autumn 1931, both Hawks were purch-ased with dollars from government funds, their transfer being approved by Göring. Tr.*]

Centre: *Some notable German aviation personalities at the 1936 air show in Oberwiesenfeld. From left: Oberst Udet, Hanna Reitsch, Willy Stör, Peter Riedl and Major Hailer. Ernst Udet had given in to Göring's pressure and joined the new Luftwaffe as an 'Officer for special duties' on 1 June 1936.*

Bottom: *The RLM (Reichsluftfahrtminis-terium, State Air Ministry) was built in Berlin in 1934–36 to the design of Prof. Dr Sagebiel.*

Top: *Until the beginning of the war, the seaplane pilots learnt their skills on the Heinkel He 42, a stable biplane. Their training schools were situated at Pütnitz, Warnemünde and Stettin.*

Centre: *Twin-engined Heinkel He 59B floatplanes of the DVS branch at List/Sylt in 1934. Hidden under the all-embracing acronym DVS, in 1935 this formation simply changed its designation to Seefliegerstaffel 1.(Mz)/186 (see Appendix), keeping its He 59Bs.*

Bottom: *Early in 1935, civil registrations were still displayed on aircraft flown by* Fliegergruppe Münster, *the later* Aufklärungsgruppe 214. *As long as the clandestine regulations were in force no guns were fitted, but the aircraft were quite easy to rearm afterwards.*

Top: *The twin-engined Do 11 was never taken into service by Luftwaffe bomber formations. The mechanism of its retractable undercarriage gave endless trouble so that in the end the aircraft were relegated to training schools where they were flown with their wheels locked in the 'down' position.*

Bottom left: *Fitted with a retractable ventral 'dustbin' gun position and a vertical bomb magazine in the fuselage, the Ju 52/3m became the first makeshift bomber of the new Luftwaffe.*

Bottom right: *Of similar layout to the Do 11, the Do 23 was powered by two BMW inline engines and had a fixed undercarriage. It replaced the Ju 52/3m makeshift bomber in Luftwaffe formations. This Do 23 with the tactical markings 53+A13 belonged to 3./KG 155 based at Giebelstadt early in 1936.*

Top right: *A retractable ventral 'dustbin'-type gun position with one 7.92 mm MG 15 was also intended to protect the crew of the Ju 86 bomber against fighter attacks from rear below – a rather draughty affair for the gunner.*

Top left: *Production of the Ju 86A, the first modern German medium bomber, began at Junkers early in 1936. Soon afterwards, the assembly line was switched over during manufacture to the aerodynamically improved Ju86D.*

Bottom: *A Ju 86A of KG 254 during a parade flight. The Jumo 205 diesel engines did not take kindly to the frequent changes in throttle settings required during formation flying.*

Top: *The twin control columns and instrument panel of the Dornier Do 17V–2. Technological advances had to be paid for by concentrating the instruments and control levers in the smallest possible space.*

Centre: *Forerunner of a new-generation bomber, the Do 17V–2. Owing to its superior speed compared to contemporary fighters it was originally intended that no armament should be fitted.*

Bottom: *The final product, the Do 17, the 'Flying pencil'. Its rounded, glazed fuselage nose section and a defensive gun position behind the cockpit cost some velocity but despite that its maximum speed of 340 km/h (210 mph) was amazingly fast for a bomber. Do 17E–1s of II/KG 155 based at Ansbach.*

Top: *The Ju 87 Stuka was developed in Germany based on experimental flights carried out with the Junkers K–47 two-seater in the late 1920s. Ju 87A dive bombers of 2./St.G 165 are seen here at Pocking airfield, near the Austrian border, during Operation 'Otto', the German takeover of that country. [These diving tests with the K–47 were flown secretly at Lipetsk in the Soviet Union, but mainly in Sweden in 1931–33, and helped development of the structure, dive brakes, dive-bombing sights and dive-bombing techniques. Tr.]*

Centre: *A squadron of Do 17E bombers of I/KG 155 overflying the crowded airfield at Vienna-Aspern on 13 March 1938.*

Bottom: *A Ju 87A dive bomber in a landing turn towards its new base at Graz-Thalethof in Austria. Early in 1938 I/St.G 167 was transferred from Lübeck-Blankensee to the annexed Austrian territory henceforward known as 'Ostmark'. Simultaneously, the unit was redesignated I/St.G 168.*

35

Top: *Only a small number of the He 114A floatplanes joined the German naval aviation squadrons. On the other hand the He 114B export version brought in useful foreign exchange from Sweden and Romania. The photograph shows three He 114A floatplanes of 1./506 coastal squadron based at Dievenow.*

Centre: *Some of the aircraft that were experimentally built for the Luftwaffe were later exported. After the Arado Ar 95A floatplane had not proved to have any significant advantages compared to the He 60 it was intended to replace, three of the nine aircraft completed were sold to Chile.*

Bottom: *The indestructible He 60 remained in naval aviation service as shipboard and coastal reconnaissance aircraft until the introduction of the Arado Ar 196. A crane puts the He 60 60+B31 of 1./306 coastal squadron in its element in September 1936.*

36

Top: *The Hs 123, known in service slang as 'Obergefreiter' (Senior Lance Corporal) was delivered to Luftwaffe dive bomber units as predecessor of the Ju 87 Stuka. The aircraft park of I./St.G 165 at Kitzingen in 1937 also consisted of these portly single-engined sesquiplanes.*

Centre: *The He 51 fighter was already below international standards late in 1936 and for that reason was released for delivery to the Spanish Nationalists via the Condor Legion. Only a limited number of He 51s remained in service in Luftwaffe fighter training schools.*

Bottom: *The Ar 68 replaced the He 51 in many fighter formations before the Messerschmitt Bf 109 was accepted for series production. One of the units so equipped until summer 1938 was I/JG 334 based at Wiesbaden-Erbenheim, the illustration showing Ar 68E fighters of its 3.Staffel.*

37

Top: *The two Oberleutnants Hannes Trautloft and Max Ibel follow with expert eyes the flight behaviour of the new Messerschmitt Bf 109D fighter at Bad Aibling in 1937. Only a few years later, both were to gain fame as exceptional fighter pilots and leaders.*

Centre: *II/JG 234, based at Düsseldorf, received its first Bf 109D fighters in June 1938. The re-equipment of parent formations in Germany caused occasional supply bottlenecks for the three German fighter squadrons of the Condor Legion in Spain.*

Bottom: *A parade line-up of Bf 109D fighters at Jüterbog-Damm. During the friendly visit by the Yugoslav Prime Minister on 18 January 1938 II/JG 132 'Richthofen' demonstrated its new 'wonder-fighters' as proof of the combat-readiness of the Luftwaffe.*

Modern Czech-built border fortifications gave the German leadership cause for serious reflection while planning 'Operation Green' and possible resistance during the annexation of Sudentenland was taken into account. However, on the day of action these fortifications turned out to be only half-completed and unarmed.

Top: *Due to the unavailability of really suitable aircraft types most long-range reconnaissance* Staffeln *were initially equipped with the He 45 biplane, such as 50+A13. This aircraft belonged to 3.(F)/125 based at Würzburg, and is seen outside the hangars at Galgenberg in March 1937.*

Centre: *During the critical early-October days when the German forces moved into Sudentenland 1.(F)/24 was deployed at an advanced landing ground at Weiden near the Czech border. Alarm was raised when a Do 17F long-range reconnaissance aircraft had to make a one-wheel landing, but the airfield ambulance was not needed: nobody was hurt.*

Bottom: *A Do 17F crew taxi their aircraft from its parking place near the edge of the woods to the grass take-off runway at the Weiden advanced landing ground. The regular base of 1.(F)/24 at that time was Kassel-Rothwesten.*

Top: *From 1937 onwards the re-equipment programme of Luftwaffe bomber formations was running in high gear. The majority of the units received various versions of the He 111, by then the standard German medium bomber. Thus, in 1938, II/KG 355 based at Giebelstadt was flying the He 111J powered by DB 600CG engines.*

Centre: *For inexplicable reasons the long-range bombers (the so-called 'Ural bombers') were deleted from the development programme by the RLM. It was believed that a tactical bombing war could be fought with a fleet of medium bombers. The first prototype of the four-engined Do 19 in flight.*

Bottom: *The Junkers Ju 89 could have been for the Luftwaffe what the Boeing B–17 Flying Fortress and the Consolidated B–24 Liberator were later for the Americans, but did not advance beyond the construction of two prototypes. This illustration shows the Ju 89V–2 with a retouched-in dorsal turret.*

Top left:
Launching of the aircraft carrier Graf Zeppelin *at Kiel on 8 December 1938. Designated 'Carrier A' in the naval construction programme the* Graf Zeppelin *was to have been commissioned in February 1940 but military events overtook planning. As a result, the completion and fitting-out of the ship were postponed.*

Top right: Gen Maj *Helmuth Förster (right), commanding General of the Luftwaffe* Lehrdivision, *presents the units of his command to the 'Fuhrer' and Generalfeld-marschall Göring at Barth near the Baltic Sea in 1939.*

Bottom: *The Berlin–Rome Axis, which came into being in October 1936, marked the start of a period of continuous fact-finding visits exchanged by the military of both countries. This photograph records the arrival in Berlin-Staaken of the Italian Air Marshal Valle on 24 June 1939. He was received by Generaloberst Milch accompanied by generals Stumpf and Kesselring.*

3. The Condor Legion in Spain

The intervention by elements of the German armed forces in the Spanish Civil War had a prelude that is not generally known.

While flying the South Atlantic route between Bathurst and Las Palmas on 20 July 1936, the Lufthansa Ju 52/3m D–APOK was suddenly requisitioned for an undetermined period at its terminal on the Gran Canarias by the local military commander and, despite protests by the Lufthansa authorites, impressed in Spanish service to drop leaflets. Two days later, the Ju 52/3m – which in the meantime had been chartered – took the military governor of the island, General Luis Organz, and two other officials to Tetuan in Spanish Morocco, one of the two focal points of the Spanish military rebellion. There General Francisco Franco had succeeded in winning over the entire garrison to the National Spanish cause against the People's Government ruling in Madrid. In the northern part of the Spanish motherland there was General Mola, in the southern half General Qeuipo de Llano; both of them were assembling cadres for further action taken from the troops which had risen against the Socialist–Communist regime.

It was essential now to get the well-trained and well-equipped Moroccan units from North Africa across the Straits of Gibraltar to the Spanish mainland to unite them with the local pro-Nationalist troops. However, as almost the entire Spanish Navy had remained on the government side, sea transport was out of the question. The only possibility was an 'air bridge', for which the Nationalists lacked suitable transport aircraft. In this situation, General Franco appealed for help to the young and similarly nationalist governments of Italy and Germany, requesting large-capacity transport aircraft to carry the 'Morros' to the Iberian peninsula. In his request he appealed to the solidarity of both countries to counter the attempt by the Soviet Union to establish a Bolshevik bastion in Spain.

During a quickly summoned conference on 25 July 1936, Hitler, in the presence of Göring, the Minister of War von Blomberg and a representative of the German Admiralty gave his permision to the Spanish commission for an immediate transfer of 20 Ju 52/3m transport aircraft with their crews. Exactly on the day these aircraft arrived in Tetuan, the Italians allocated 11 Savoia-Marchetti SM.81 transports for the same task.

The German assistance to Spain was handled by a firm named HISMA (Compania Hispano-Marroqui de Transportes Tetuan), while a 'Sonderstab W' ('Special staff W') under Gen Ltn Helmuth Wilberg, established late in July 1936, was responsible for the personnel aspects of the German assistance.

The 20 Ju 52/3m transports which arrived in Spain by air were followed early in August by the second consignment of German assistance aboard the steamer Usaramo. This consisted of twenty 20 mm light anti-aircraft guns, six He 51 biplane fighters and about 100 tons of other war material. In addition to the crew, the Usaramo also carried 86 German volunteers, including ten transport aircraft crews and six fighter pilots. These were intended to provide fighter escort for the transport aircraft between Tetuan and Sevilla, but were ordered to avoid all other operational tasks.

However, as the constantly overloaded Ju 52/3m transports managed to ply there and back three to four times a day without any interference from Republican fighters the six He 51s were available to train Spanish fighter pilots and, late in August 1936, were finally transferred to a Spanish fighter squadron.

The first time German airmen were involved in fighting happened on 13 August 1936 when two Ju 52/3m transports, converted into makeshift bombers and flown by the crews of Hptm von Moreau and Oblt Graf (Count) Hoyos bombed and badly damaged the Republican armoured cruiser Jaime I which had opened anti-aircraft fire on Ju 52/3m transports flying to and from Sevilla.

In the meantime, the volunteer German

fighter pilots sat around doing nothing and had to look on as their Spanish colleagues, still less than familiar with the German He 51 fighters, crashed them one after another.

Urged by the commander of the German detachment, *Major* Scheele, General Franco and the RLM in Berlin gave their agreement that in future German fighters could also be flown by German pilots. After several more Ju 52/3m transports had been converted into makeshift bombers by fitting vertical bomb magazines, two Ju 52/3m-*Ketten* (flights) known as the 'Pedros' (led by *Hptm* v. Moreau) and 'Pablos' (*Oblt* Rudolf Jöster) began to fly operational sorties with German aircrews, bombing Republican targets in the Madrid-Guadalajara area. Apart from that, the transport flights between Tetuan and Seville remained the main task of the Ju 52/m detachment until mid-October 1936. By that time no less than 13,500 troops and 269 tons of war material had been flown across to the Spanish mainland.

Until the end of October 1936 all German assistance shipments and other activities had to take place in the strictest secrecy. All participants of the '*Unternehmen Feuerzauber*' ('Operation Fire Magic'), the cover name for the whole action, were formally discharged from the German armed forces before they went to Spain, and wore civilian clothes on the way out. However, with the appearance of the first Soviet aircraft, flown by Soviet aircrews early in November 1936, as well as the increasing use of the 'International brigades' on the Republican side, came the formation and official establishment of the German auxiliary corps 'Condor Legion', organized under the cover name of '*Ubung Rügen*' ('Rügen Exercise'). Together with its first commander, *Gen. Maj* Hugo Sperrle, uniformed members of the German armed forces landed in Spain for the first time on 15 November 1936. By far the predominant part in the 'Rügen Exercise' was played by Luftwaffe formations which eventually had the strength of a *Fliegerkorps* and consisted of the following:

A/88 Reconnaissance *Staffel*
AS/88 Seaplane reconnaissance *Staffel*
B/88 Base-airfield operating company at the aircraft park
F/88 Anti-aircraft detachment (four heavy, two light AA batteries)
J/88 Fighter *Gruppe* (three, sometimes four *Staffeln*)
K/88 Bomber *Gruppe* (as above)
Laz/88 Field hospital
Ln/88 Signals detachment
MA/88 Ammunition depot
P/88 Depot and aircraft supply detachment
San/88 Medical detachment
VB/88 Experimental bomber *Staffel*
VJ/88 Experimental fighter *Staffel*
VS/88 Liaison staff to the Spanish and Italian Air Forces
W/88 Weather station
S/88 Operations staff

The numerically much smaller German Army and Navy units operated under the covernames of '*Imker*' and '*Nordsee*' ('Apiarist' and 'North Sea').

The first to report his formation in operational readiness was *Major* Fuchs, *Gruppenkommandeur* of K/88: the Ju 52/3m bombers had flown direct from Greifswald in Germany via Rome and Melilla to Seville; aircraft to the other detachments first had to be unloaded and assembled. These comprised He 51 fighters for J/88, He 70 low-wing and He 46 high-wing monoplanes for A/88 and He 59 and He 60 floatplanes for AS/88, on the way by ship. More modern fighters, such as the first He 112 and Bf 109 were to appear in VJ/88 at Sevilla-Tablada early in December 1936.

The 88 mm Flak battery under *Hptm.* Aldinger took over the anti-aircraft defence of Sevilla-Tablada, the base of P/88, and the transfer and assembly centre for German aircraft delivered to Spain. It was from Tablada that the Condor Legion bombers flew their first raids on Republican ports along the Mediterranean coast and airfields in the Madrid area. But not for long: due to the massed Republican fighter defences these bombing sorties very soon had to be switched to the evening and night hours. The fighter pilots too had an uphill struggle; despite the first aerial victories they very quickly had to recognize the inferiority of their He 51 biplanes compared to the Soviet I–15 'Chato' and I–16 'Rata' fighters which had been shipped aboard Soviet freighters to Cartagena, Alicante and Barcelona, replacing the obsolete Republican Dewoitines and Nieuports. The Fiat CR.32 biplane flown by the Italian volunteer squadrons gave a better account of itself: with this highly manoeuvrable aircraft it was possible to escape numerically superior opposition either by diving away or by a steep downwards turn. The situation was even worse with the bombers: the obsolete Republican Potez 54 and Fokker XVIII's had been replaced by the fast Soviet Tupolev SB–2 'Katiuska' bombers, which left the He 51 standing.

During the so-called 'Winter Battle for Madrid' the Nationalist flying formations were moved closer to their targets. The German bombers transferred to Salamanca, the fighters to Escalona, Almarox and Avila, and one *Staffel* to Leon, on the Northern Front. The Condor Legion reconnaissance aircraft changed their operational bases very frequently, being deployed according to the situation. On the other hand, the seaplanes of AS/88 remained based at Melilla in the Spanish Morocco until February 1937.

During the winter months of 1936/37, the He 51 fighters already had to be used more and more frequently as ground support and attack aircraft, a role that became their staple diet after the arrival of the first Bf 109 fighters in March 1937. In this hard work the He 51s were assisted by the few He 123 ground attack aircraft shipped to Spain.

It came as a relief to the Ju 52/3m bomber crews when they were promised the delivery of the first 'real' bombers, the Ju 86, Do 17 and He 111, beginning January/February 1937. However, only four machines of each type were to be formed into VB/88 in order to work out the best attack and ground-control tactics based on combat experience.

Then came the battle of Guadalajara in mid-March 1937 when the Italian Legion suffered heavy losses, and this, together with the stagnating Nationalist advance on Madrid, moved the Nationalist command to change direction and take the strip of Basque territory along the Bay of Biscay still in the Republican hands to end the need to fight on two fronts.

The entire Condor Legion was moved north to become the main Nationalist air weapon in that area. Simultaneously P/88 was transferred to Leon in Northern Spain and shipments of German aircraft now went to Vigo instead of Cadiz.

When the offensive began in the Northern Sector on 31 March 1937 directed towards Ochandiano-Bilbao, the Condor Legion and its subordinated Italian squadrons were faced by only weak Republican flying units. The crews were specifically ordered, in the interests of uninterrupted advance, to disregard civilian population when attacking the road and rail communications systems in the enemy's rear areas – which led to the bombing of Ochandiano, Elgueta and Guernica, which still stains the name of the Luftwaffe. Of these, the bombing of Guernica on 26 April 1937 will probably be held against the Luftwaffe for ever. In bad visibility and blinded by enormous smoke and dust clouds thrown up by the first bombers the following sections dropped their bombs without a clear view of the target, right into the inferno. The actual targets, a bridge and a road fork hard east of the town, were not hit by a single bomb. It is true that the originally publicized figure of 1600 dead later

proved to be exaggerated, but even the officially established figure of just over 300 in no way justifies such a badly planned and executed bombing raid.

In the operations that led to the capture of Bilbao the Condor Legion's VB/88 with its new He 111, Do 17 and Ju 86 medium bombers was joined by the first Bf 109s which equipped 2.J/88 led by *Oblt* Lützow, and immediately proved far superior to the Republican fighters.

The 'Iron ring' around Bilbao was broken on 12 June, and the large coastal town itself was taken seven days later. The rest of June was spent attacking towards Santander but none of these attempts had the right driving force.

On 6 July 1937 came a Republican relief offensive from the surrounded Madrid area. The attacking divisions were reinforced by the XI and XIII International Brigades and drove north, towards Brunete and Quijorna.

Leaving behind only small sections of its units, the Condor Legion transferred back to its old airfields at Avila, Villa del Prado, Escalona and Salamanca to support the Nationalist forces as 'flying artillery' in this area of Republican penetration, as well as to protect them against attacks by the 'Katiuskas' and 'Natashas' [Soviet-built R–5/R–Z close-support biplanes, *Tr.*]. The battle lasted till the end of July when the situation finally stabilized around Brunete. This dogged combat in searing heat, in which the air forces played a significant part with their 'rolling' bombing attacks, cost some 30,000 casualties on both sides.

Early in August, the Condor Legion squadrons were once again deployed along the Northern Front to bomb free the way for the ground troops attacking the coastal region from Bilbao to Gijon. In this, the HE 51s,* named 'Trabajadores' ('Workers' or 'Labourers') by the Spaniards, had the task of neutralizing the Republican road and rail communications in low-level attacks (3.J/88). The other two *Staffeln* (1. and 2.J/88) were already equipped with the Bf 109 and were used exclusively as fighters. By then, almost the entire bomber *Gruppe* (K/88) was already equipped with the He 111s, and the Do 17s of A/88 were also used as bombers in raids on targets in the Republican hinterland. Gijon, the most westerly town in the Republican territory, was taken on 21 October 1937 after aerial superiority had been decisively won by the Condor Legion and its subordinated Italian and Spanish squadrons.

*A squadron of He 51s were later also known as 'Cadenas' ('Chain links') from the way they operated in repeated low-level attacks. *Tr.*

The removal of this threat from the Basque region and Asturia allowed the Nationalist command to concentrate on taking Madrid again. Beginning early December 1937 the Italian and German bomber squadrons started carrying out raids on the Republican airfields around Madrid, Bujaraloz, Guadalajara, Alcala de Henares and others to eliminate the opposing air threat. But the Republican defences were alert, and the bombers were nearly always hard-pressed by swarms of Soviet-built I–16 and I–15 fighters.

Then a surprise pincer movement by Republican troops broke through the Nationalist lines and surrounded the exposed town of Teruel on 16 December 1937, once again crippling the Nationalist advance on Madrid. The new commander of the Legion, *Gen. Ltn.* Volkmann, was forced to use his flying formations against the Republican troops ringing Teruel, and so Calamocha became the most advanced operational airfield used by the German aircraft. In icy cold, with temperatures dropping to −17°C, fighters of J/88 took off from Calamocha to carry out low-level attacks and fly escort for Nationalist bombers around Teruel. The longer-ranging bombers were based at Zaragoza, El Burgo de Osma, Alfaro and Bunuel but their raids on the encircling Republican brigades could not prevent them taking the town on 8 January 1938.

Right through January and almost the whole of February, Teruel remained the main target of the 'Legion'. A new type of aircraft appeared over the battlefield in mid-February, the first Ju 87A Stukas of the '*Jolanthe-Kette*' (Flight), attacking selected point targets. It was in January and February 1938 that, thanks to flying several sorties a day, the German fighter pilots *Ofw* Seiler and *Oblt* Balthasar scored their impressive series of aerial victories.

Reduced to a heap of ruins, Teruel was finally retaken by the Nationalist troops on 21 February 1938. The old frontline was re-established – a victory purely in prestige terms, without any strategic advantages.

Taking into consideration the situation on the Madrid front, the 'Caudillo' decided on a new distant target: breakthrough towards the east along the Ebro river, advancing right to the Mediterranean coast. This new attack began on 9 March 1938. Following the already established recipe, aircraft of the German and Italian 'Legions' were given the task of destroying Republican airfields, communications and bridges, the annihilation of the enemy air force on the ground and interference with their supplies to the front. The He 59 seaplanes of As/88 based at Pollensa/Mallorca also took part in this operation, carrying out nocturnal bombing raids on Republican railways near the coast and the ports of Valencia, Castellon and Tarragona.

Assisted by continuous low-level attacks by Bf 109 and He 51–*Staffeln*, the Nationalist forces managed an uninterrupted advance to Caspe on the Ebro river, reaching it on 17 March 1938, before the Republican resistance began to stiffen, necessitating an advance on a broader front. In this, the 'Condor Legion' undertook the support of the Northern Army, while the National Spanish aviation formations operated in the central sector, and the Italian aircraft along the southern flank.

By 15 April 1938, the day when Nationalist troops led by General Aranda reached the Mediterranean coast at Vinaroz, *Hptm* Harder had increased the number of his aerial victories to ten, while *Fw* Boddem and *Lt* Seiler had scored nine confirmed victories each.

With a swing southwards, the Nationalist advance then turned towards Valencia, which necessitated another transfer of the flying formations. From then on, La Cenia became the new base for the Condor Legion fighters and the He 45 flight. The anti-aircraft defences were looked after by F/88, although three 88 mm Flak batteries remained with the advancing Spanish corps because the effectiveness of these guns in direct fire against the Republican ground fortifications made these fully-motorized batteries indispensable at the front.

However, due to the bad weather and lack of drive the advance was rather slow in May 1938. Furthermore, supplies of new and replacement aircraft from Germany had dried up: it was the time of Germany's entry into Austria and the Sudetenland, which required all available forces at home. On the Spanish front, this was reflected in a drop in the operational readiness rate of German formations. It was not until June and July 1938 that the Nationalist forces managed to overcome the mountainous foreland around Valencia, opening the way for an attack on the port itself.

Although the prepared Communist positions along the northern shores of Ebro, at the foot of Sierra Montsant, had been recognized and reported by reconnaissance aircraft of A/88 in mid-July, the Republican relief offensive on 25 July 1938 came as a complete surprise to the Nationalists. Consequently the Republican forces managed to cross the Ebro and establish several bridgeheads within three days, followed by an advance that gained them about 600 sq km of territory south of the river.

The number of victories achieved by German fighter pilots and the combat reports of other German aircrews between July and mid-November 1938 reflect the hard struggle in the air and on the ground during this period. Enormous bomb loads were dropped on the Ebro crossings and the fortified positions of the Republican divisions which had broken into the Nationalist lines. The 22 Bf 109 fighters which arrived from Germany on 15 July 1938 came just in time to combat the day and night bombing raids by Soviet-built SB-2 'Katiuska' bombers. This new delivery also meant that 3.J/88 could be equipped with the new fighter.

By 16 November 1938, the Nationalist troops, with the utmost effort, had finally managed to retake the whole area occupied by the Republican attack. During this period the Republicans lost over 300 aircraft, a loss that broke the backbone of their air force. It was during these months that German fighter pilots like Werner Mölders, Otto Bertram and Walter Grabmann were able to achieve most of their aerial victories. They were also more in the public eye than the bombers, reconnaissance and naval aviation crews who despite their daily, hard, operational sorties remained the silent 'water carriers' of the Legion.

There was hardly any effective Republican resistance to the Nationalist drive that took Catalonia right up to the Pyrenees and the French frontier. At that time, it was mainly a matter of interdiction bombing of the roads leading north, which were overflowing with dispirited republican divisions and International Brigades retreating towards France, and securing the Southern Front near Valencia against another possible sudden Republican attack.

The Soviets had long since withdrawn their pilots and stopped deliveries of further war material and, although the Spanish pilots supporting the Republican government tried to provide a makeshift protective umbrella over their demoralized ground forces, their efforts were largely in vain. In terms of quantity alone, they could not hope to offer any serious resistance to the National Spanish, Italian and German squadrons.

Fighting in Spain came to an end on 28 March 1939. The occupation of territories still in Republican hands, including the capital Madrid, remained a matter for the Spaniards themselves.

The Germans left nearly all their aircraft and other war material in Spain. On 22 May 1939 the Condor Legion paraded before General Franco and its last commander, *General-major* von Richthofen, and just four days later

the German personnel started to board passenger ships of the KdF fleet* at Vigo, destination Hamburg.

After the award of decorations all participants in the 'Rügen Exercise' took part in the 'Great Parade' in Berlin on 6 June 1939, marching past Adolf Hitler and the leading personalities of the National Socialist Party and the Armed Forces. A total of 298 Legionnaires did not see this great occasion: they had lost their lives in Spain.

The Luftwaffe gained in many respects important experience in the Spanish Civil War which would soon prove to give them definite tactical advantages in the European conflict, only a few months away. Among other innovations was the 'loose' combat formation of smaller fighter units, such as '*Rotten*' (pairs) and '*Schwärme*' (flights) which had considerable advantages compared to the 'closed' flight, squadron V-formations or 'follow the leader' squadron columns, standard practice until that time. This new combat formation could only be forged in actual combat, as it happened in Spain (see note to illustrations on page 61). Then there was the use of the He 51 biplanes in ground attack and close support, which provided important lessons regarding such tactics.

However, the fateful use of medium bombers as 'flying artillery' in Spain was to have far-reaching consequences in the future, when the same tactics were attempted on a much wider scale. It also shelved planning for a strategic bomber force, which was to remain completely neglected.

On the other hand, the experiences gained in Spain by the few Stuka crews were to be of great significance once they had been evaluated and assimilated by the parent formations. The value of this Spanish experience was proved many times over during the first two years of the Second World War.

The Spanish Civil War also provided many almost 'textbook' situations to test the heavy and light anti-aircraft guns in action against flying and ground targets, which allowed comparison and choice in optimum combat methods. Valuable experiences during the 'Rügen Exercise' were also gained by the signals companies, be it in the operational evaluation of their radio sets, improvements in ground-to-air communications under combat conditions or progress in the air traffic control. All these experiences found their place in the various

*KdF = *Kraft durch Freunde* (Strength through Joy), a German Labour Front organization specially formed to enable ordinary working people to go on cruises abroad. Apart from gaining more support for the 'Führer', it also had tremendous propaganda value. *Tr.*

service and field regulations issued during the following months and years.

All told, about 14,000 German soldiers passed through the 'cadre school' of the Spanish Civil War, although at no time there were more than 6500 men in action at the sae time on the Iberian peninsula.

Top: *Moroccan troops awaiting their air transport at Tetuan in Spanish Morocco. The German leadership provided the Spanish Nationalists with 20 Ju 52/3m aircraft to transport these well-trained formations to the Spanish mainland.*

Bottom: *Mourning guard of honour for the first German airmen to lose their lives in Spain. Uffz Helmut Schulze was killed together with his flight mechanic Uffz Zech when their Ju 52/3m crashed at Jerez de la Frontera on 15 August 1936.*

The central base of the German voluntary formations was the Seville-Tablada airfield. It also accommodated the Condor Legion aircaft park. Sèville-Tablada was protected by an 88 mm Flak battery commanded by Oblt Hermann Aldinger, seen here being 'waited on' by one of his men.

Top: *The Ju 52/3m transports forming the so-called 'Pablo's Flight' were commanded by* Oblt *Rudolf Joester (left) assisted by his navigator* Oblt *Piecha. These transport aircraft were retroactively fitted with bomb magazines and in late summer 1936 used to raid targets in the Madrid and Sierra Guadarama areas.*

Centre: *The squadron commander of 3.K/88,* Hptm *Krafft von Dellmensingen had mounted two fixed MG 17 machine guns atop the inner mainplanes of his Ju 52/3m to protect it against fighter attacks from dead ahead. These 'scare weapons' fully justified themselves in action.*

Bottom: *After the arrival of the Condor Legion in November 1936 three complete Ju 52/3m squadrons were operating as* Kampfgruppe *88 (K/88) against the Republican-held Mediterranean ports. An aircraft of 3.K/88 over the Sierra de Ronda mountain range.*

51

Top left: Hptm
d.Reserve *Josef
Veltjens, a Pour-le-
Mérite airman of
the First World
War, was put
forward by the
German government
as a 'dummy' for
supposedly private
armament deliveries
to Spain. His role
became superfluous
after the official
despatch of the
Condor Legion.*
Top right: *Among
the first volunteer
fighter pilots in
Spain was Oblt
Herwig Knüppel
who returned to
Germany with eight
confirmed aerial
victories. He was
killed in action
over France as
commander of II/JG
26 on 19 May 1940.*

Bottom: *Personnel
of 1.K/88 pay a visit
to their comrades of
the Flak artillery at
the Talavera front.
On extreme right is
Uffz (Cabo) Richard
Schade, 2nd pilot of
a Ju 52/3m.*

Top: *In spring 1937 1.J/88 under* Staffelkapitän *Palm was based with its He 51s at Escalona on the Madrid front. Note the Condor Legion number plate on the Mercedes-Benz fuel bowser.*

Centre: *In mid-April 1937 the pilots of 1.J/88 were longing to fly the promised Bf 109 fighters when the squadron was taken over by* Oblt *Harro Harder (right). By that time 3.J/88 commanded by* Oblt *Douglas Pitcairn (left) were already using their He 51s in the ground attack role.*

Bottom: *With his 12 credited aerial victories Wolfgang Schellmann, known as 'Gedalje', was the second-highest-scoring German fighter pilot in Spain next to* Hptm *Werner Mölders (14).* Hptm *Schellmann led the 1.J/88 from 19 December 1937 to 2 September 1938.*

Top: *After the aircraft park of the Condor Legion had been transferred from Sevilla to Leon in northern Spain in March 1937, all bulky aircraft components that were wider than the railway profile had to be transported by road from the port of discharge at Vigo.*

Centre: *The A/88 reconnaissance detachment of the Condor Legion had a flight of He 45 biplanes whose task was battle-zone reconnaissance and, in addition, acting as light bombers against enemy frontline positions. Here the He 45B 15.21 at Leon in mid-1937. The Spaniards had nicknamed this aircraft type 'Pavo' (Turkey).*

Bottom: *From the very beginning, the He 70 was relieved of its proper reconnaissance activities and used as a high-speed bomber. In February 1937 the He 70 flight commanded by Lt Runze carried out several successful raids against power stations in the spur of the Pyrenees mountains.*

Top: *The first modern German bombers, which arrived in Spain at the beginning of January 1937, were assembled in Seville and formed into the experimental VB/88 bomber squadron. This unit was commanded by* Hptm *Rudolf von Moreau, standing (right) in front of an early He 111B bomber. Facing him is 'Old Bengsch' and Graf (Count) von Hoyos.*

Centre: *In addition to four He 111Bs the experimental VB/88 also had four examples of the diesel-powered Ju 86D, seen here head-on. On this aircraft the undercarriage did not retract into the engine nacelles as on other bombers, but sideways and upwards into the wings.*

Bottom: *Four Do 17E 'Flying Pencils' completed the aircraft strength of VB/88, but were transferred to AS/88 reconnaissance detachment in summer 1937. The depicted 27.17 already carries the insignia of the reconnaissance squadron.*

Top: *The He 59 floatplanes of A/88 commanded by* Hptm *Harlinghausen and based at Pollensa on Majorca carried out a series of nocturnal raids on the Republican ports of discharge and railways along the Mediterranean coast. A He 59B of AS/88 armed with a flexible 20 mm cannon in the nose gun position.*

Centre: *The pilots of 2.J/88 commanded by* Oblt *Günther Lützow were the first to receive the Messerschmitt Bf 109B fighters. As a result, from March 1937 onwards the superiority hitherto enjoyed by the Republican-flown Soviet I–15 'Chato' and I–16 'Rata' fighters had come to an end. At the same time the He 51 biplane fighters began increasingly to be used in the ground-support role.*

Bottom: *A ruined section of Guernica, a small Basque town. A negligently planned bombing raid by K/88 on a railway junction in the suburbs on 26 April 1937 resulted in extensive bomb damage in the town centre and the death of more than 300 civilians.*

Top: *On 13 June 1938, during the Nationalist advance towards Valencia, aircraft of K/88 carried out an effective bombing raid on railway installations at Nules. The escorting fighters shot down eight Republican aircraft but, for all that, six He 111s returned to their base at Alfaro with extensive battle damage.*

Bottom: *A group of Nationalist Spanish officers of the 7th Division during the Ebro battles in August 1938. The divisional commander Coronel Lopez Bravo (centre) is seen discussing aerial reconnaissance over the battle zone with Major Graf (Count) Kerssenbrock (4th from left), commander of A/88.*

57

Top: *The Tupolev SB–2 bomber supplied to the Republicans was an excellent combat aircraft. Known in Spain as 'Katiuska', it was far superior to the sedate Ju 52/3m on the Nationalist side.*

Centre: *The Prussians are not prudish. Anybody who could not find the 'thunderbox' at Alfaro must have been blind: the inscription on this elegant outdoor convenience reads, 'Hooray! Here shits the 1st Company!'*

Bottom: *It was not until the arrival of the He 111 that 'Condor Legion' bomber squadrons could reach parity with the opposition. And not only that: with a bomb load of 1000 kg (2205 lb) and a maximum speed of 375 km/h (233 mph) the He 111 actually proved superior to the SB–2 'Katiusca'. Note the unusual wing motif on the fuselage nose/centre section of this He 111, and the 'chimney sweep' individual insignia on the fin.*

Top: *In 1938 the central operational base of the Condor Legion was La Cenia near the coast, south of the Ebro estuary. Initially intended as a pure fighter airfield, it was also used by Stukas and reconnaissance aircraft.*

Centre: *A Bf 109D undergoing undercarriage functional tests at La Cenia. A few of the later Bf 109E fighters also managed to reach Spain near the end of hostilities. Capable of a maximum speed of 550 km/h (342 mph) the Bf 109E was 100 km/h (62 mph) faster than the D-version.*

Bottom: *Captured Soviet-built aircraft at the Valls airfield, north of Tarragona, in March 1939: the Polikarpov I–15 (known in Spain as 'Chato') and, behind it, the Polikarpov R–Z ('Natasha') close-support aircraft.*

Top: *The officer in the oversize peaked cap is Gen Maj Hugo Sperrle (covername 'Sanders') who was the first commander of the Condor Legion from early November 1936 to 31 October 1937.*

Bottom: Gen Maj *Sperrle was followed as commander of the Condor Legion by* Gen Maj *Hellmuth Volkmann (covername 'Veith'); he remained in Spain until 13 November 1938. General Volkmann (left) with General Alfredo Kindelan (centre), commander-in-chief of the National Spanish Air Force.*

Top: *On 10 September 1938* Maj *Gotthard Handrick (left) handed over command of J/88 to* Hptm *Walter Grabmann (right), who led the three German fighter squadrons in Spain until the Armistice on 28 March 1939.*

Centre: *Refuelling a Bf 109D fighter at La Cenia. Due to the lack of aerial opposition during the final months of fighting in Spain the Condor Legion pilots increasingly participated in ground battles, carrying out low-level attacks on roads, railway lines and airfields.*

Bottom: *A friendly exchange of views among Luftwaffe fighter experts. From left:* Oblt *Wolfgang Lippert,* Hptm *Werner Mölders (commander of 3.J/88),* Hptm *Siebelt Reents (commander of 1.J/88) and* Oblt *Rolf Strössner. With 14 confirmed aerial victories* Hptm *Mölders was the most successful Condor Legion fighter pilot. Later known and respected as 'Vati' (Daddy) Mölders, he originated and introduced the 'two pairs' or 'finger-four' fighter combat formation which still holds good today.*

61

Top left: *Maintenance work on a Ju 87B under the most primitive conditions. Altogether eight Ju 87 Stukas of both A and B versions were sent to Spain, this particular aircraft being received in October 1938. These Ju 87s were first attached to J/88 and then, from late 1938, to K/88.*

Right: *Lt Siegfried Trogemann was one of the Stuka pilots of the 'Jolanthe' Kette (flight) who tested the dive bombing tactics in Spain. Note the traditional 'Jolanthe' insignia on the undercarriage fairing of his Ju 87B.*

Bottom: *The entrance to the barracks and tent camp at Döberitz near Berlin. All 'Legionnaires' who had taken part in the fighting in Spain were assembled here to participate in the great homecoming parade in Berlin on 6 June 1939.*

4. The Polish Campaign

If Hitler's *Lebensraum-Politik* involving the annexation of Austria and the Czech territories had yet not led to any threatening reactions from the Western powers, France and Britain, the crushing and occupation of Poland was overnight to fan the smouldering possibility of a European war into a world-wide conflagration. To be sure, by lightning diplomatic action in signing the non-aggression pact with the Soviet Union on 23 August 1939 the German leadership had protected itself against the intervention of that power bloc, but the hope that France and Britain might possibly take their assistance pact with Poland only half-heartedly was to prove false.

In '*Fall Weiss*' (Case White), cover-name for military preparations against Poland, the cue for the Luftwaffe was '*Ostmarkflug*' ('Eastern border-country flight'). That key word was signalled to the Luftflotten on 26 August 1939 who in turn immediately ordered all their subordinate units into combat-readiness, including the transfer of all flying formations to their allocated operational bases. Numerically and structurally the Luftwaffe was absolutely superior to its Polish opponent but it would certainly have been a fiasco if Britain and France had followed their declaration of war on 3 September 1939 with immediate aerial operations against Germany. The Luftwaffe could have hardly been equal to the combined air-forces of both Western nations in a two-front war.

It will probably never be possible with complete accuracy to list all aircraft classes in Luftwaffe service, but a summary of formations existing at that time gives a reasonably accurate picture. On 1 September 1939 the Luftwaffe General Staff had the following operational formations at its disposal:

29 Army (short-range) (H) reconnaissance *Staffeln*
  Hs 126, He 46, He 45, Fi 156

23 long-range (F) reconnaissance *Staffeln*
  Do 17F, Do 17P, Do 215
15 day fighter *Gruppen* (light fighters)
  Bf 109D, Bf 109E, Avia B534
1 carrier-borne fighter *Staffel*
  Bf 109D, Bf 109T
3 provisional night-fighter *Staffeln*
  Ar 68F, Bf 109D, Bf 109E
10 'destroyer' *Gruppen* (heavy fighters)
  Bf109B, Bf 109C, Bf 109D, Bf 109E, Bf 110B
30 bomber *Gruppen*
  He 111P, He 111J, Do 17Z, Ju 86D
1 bomber training/instructional *Gruppe*
  Ju 88A
9 dive bomber *Gruppen*
  Ju 87A, Ju 87B
1 carrier-borne dive bomber *Staffel*
  Ju 87B, Ju 87C
1 close support *Gruppe*
  Hs 123
5 transport *Gruppen*
  Ju 52/3m
2 shipborne *Staffeln*
  He 60, He 114, Ar 196A
5 naval aviation *Staffeln* (M) – coastal reconnaissance
  He 60, He 114
4 naval aviation *Staffeln* (F) – long-range reconnaissance
  Do 18
5 naval aviation *Staffeln* (Mz) – multi-purpose
  He 59, He 115

Such formations as the weather reconnaissance, courier and liaison *Staffeln* and units of the air-sea-rescue service were still in the process of formation at that time and could be considered as only conditionally operational.

The largest part of the formations listed above came under Luftflotte 1 (General Kesselring) and Luftflotte 4 (General Löhr), being deployed in the Pomerania-West Prussia and Silesia assembly areas, while the units based in East Prussia (KG 3, St.G 1, JG 21, 1.(F)/121) together with the Luftwaffen-*Lehrdivision* (training/instructional formation comprising LG 1 and later LG 2) operated from this 'aircraft carrier East Prussia' in support of the Third German Army. Other bomber *Gruppen* flew operational sorties from their more distant home bases in Germany, such as I/KG 27 from Hannover-Langenhagen.

With the aim of achieving aerial superiority as quickly as possible, the Luftwaffe target catalogue for the first two days of the war comprised enemy airfields near the border area as well as deeper inland. To this end, on 1 and 2 September 1939 German bombs rained down on the Polish air bases at Warsaw-Okecie, Lida, Kolo, Radom, Graudenz, Thorn (Torun), Plock, Posen (Poznan), Krakau and others, but causing only minimal damage to the Polish Air Force on the ground. Most of the main Polish air bases were empty: during the last days of August the operational aircraft had been quietly moved to well-camouflaged field bases. At the same time, the manoeuvrable but rather slow and poorly armed Polish fighters were an easy prey to German escort fighters. From 3 September, the main effort of Luftwaffe bombing raids was transferred to road and rail targets, indicating a kind of strategic air warfare.

On the other hand, the Stuka and some of the participating *Zerstörer* ('destroyer' or heavy fighter) *Gruppen* operated in a purely tactical way from the very first day, flying in direct support of the northern and southern Army Groups. After the conclusive achievement of aerial superiority and the destruction of point targets such as bridges, railway junctions, stations and fortifications, the Ju 87 Stuka became the (surely overrated) symbol of the 'Blitzkrieg'. Its occasional lack of accuracy in dive bombing was easily made up for by the fearful spectacle of a Stuka-*Staffel* or *Gruppe* diving down to attack with the infernal howl of their sirens.

After the first week the campaign in Poland changed increasingly into a series of encirclement and destruction battles against the retreating Polish troops assembling in forests off the main roads – which had become the regular targets of Stukas and the heavy fighters.

By the middle of September the majority of the remaining Polish military aircraft had flown to the airfields in neutral Romania so that by 17 September those Polish aircraft still serviceable had been temporarily saved from the invasion. For that reason, the remaining P–23 and P–37 light bombers had to fly their occasional and costly attacks against the advancing German divisions without any fighter protection. After the air war in Poland had taken this turn, the Luftwaffe High Command began to withdraw individual *Gruppen* back to Germany beginning mid-September, transferring some to protect the Western areas. The remaining *Staffeln* did not encounter any more opposition in the air and suffered only a few more losses due to Polish anti-aircraft guns and infantry weapons.

In the meantime the capital Warsaw and the fortress of Modlin became the most prominent assembly and resistance centres in central Poland and climbed into the top-level Luftwaffe target list accordingly. As much as a week before the capital was surrounded, on 24 September 1939, the Polish town commandant was requested by parliamentarians and leaflets to surrender Warsaw without a fight, and a planned Luftwaffe bombing raid was in fact postponed. However, the Polish troops surrounded in Warsaw went on preparing for street fighting and refused any talk of giving up the town. And so came the heavy Luftwaffe bombing raid on the Polish capital on 25 September 1939. About 400 aircraft, including some Ju 52/m transports whose crews unloaded stacks of stick-type incendiary bombs by hand, made up to four sorties bombing the town centre that day. On the following day the Polish defenders of the badly damaged and burning capital showed themselves willing to give in, and the capitulation of Warsaw was signed on 27 September 1939.

Modlin with its fortified installations and casemates was a different proposition, but gave in on 29 September 1939 following several Stuka attacks. Not counting the odd casualties incurred in skirmishing east of Vistula river, which lasted until 7 October 1939, with this the Polish campaign came to an end. During this final phase, the Luftwaffe activities were limited to reconnaissance flights, mainly because the confused situation on the ground had made it impossible to carry out accurate raids on the dispersed Polish army units without endangering their own troops.

Top: *After clearance of the early morning mists on 1 September 1939 Luftwaffe formations take off to attack their allocated targets in the Polish territory. The Ju 87Bs of I/St.G 77 operated from Neudorf near Oppeln.*

Centre: *Over-sized crosses on the wings of German aircraft were intended to stop nervous anti-aircraft batteries and fighters from mistakenly shooting at their own aircraft. He 111 bombers of 2./KG 27 'Boelcke' in approach flight to their target at Warsaw-Okecie.*

Bottom: *It was intended to capture the railway bridge at Dirschau intact for use by German supply transport. Although by accurate bombing the Stukas succeeded in cutting the fuse cables early in the morning of 1 September 1939 the Polish troops still managed to blow up the bridge.*

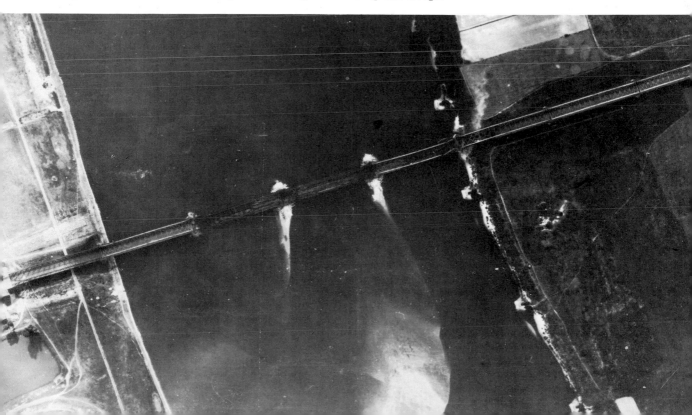

Top: *After the Anglo-French declaration of war (on 3 September 1939) their bombers were expected to fly in over the coastal area along the North Sea. This was to be prevented by a thick net of anti-aircraft batteries, some of which still operated the older type of sound detectors.*

Centre: *The Ringtrichter-Richtungshörer 36 (trumpet-type sound detector) was the standard target-acquisition equipment of Luftwaffe search-light batteries. Despite dependance on weather conditions and inaccuracies in target location these devices stayed in service to the end of the war.*

Bottom: *To protect the operational bases of the Luftwaffe along the Polish border against enemy low-level air raids they were surrounded by batteries of light and heavy anti-aircraft guns. A 20mm Flak 30 in position near the Oels airfield perimeter.*

Top: *I/KG 27 'Boelcke' was deployed to Neukuhren in East Prussia, from where it carried out bombing raids on Polish road and railway communications. The He 111P 1G + EK of* Lt Scholz, *decorated with his personal insignia.*

Below left: *Known as the 'Father of Stukas', Obstl Günther Schwartzkopff led the St.G 77 in the Polish campaign. On 14 May 1940 his Ju 87B received a direct hit from an anti-aircraft shell over Sedan and both Schwartzkopff and his gunner crashed to their death.*

Below right: *To increase the howling sound made by the Ju 87s while diving they were fitted with special noise devices (sirens) attached in front of both undercarriage fairings. However, these fittings resulted in considerable loss of speed in horizontal flight and the sirens were soon removed in some formations.*

Top: *2./ZG 1 has just taken off from its base at Mühlen on 8 September 1939 to fly escort for a Stuka-Gruppe attacking targets in the Warsaw suburb of Praga. At that time ZG 1 was still equipped with the older Bf 110B powered by Jumo 210 engines.*

Centre: *When by the middle of September the Polish fighter defences had been eliminated the Luftwaffe fighters escorting bombers often indulged in 'dog fights' among themselves. Here a Bf 109E pilot has managed to get his leader in a promising position from rear-below in his Revi (reflex) gunsight.*

Bottom: *I(Z)/LG 1 participated in the Polish campaign already equipped with the improved Bf 110C. As an instructional and demonstration formation LG 1, then based at Barth near the Baltic Sea, would always receive the latest equipment.*

Right, top: *A target-effect photograph of the railway station at Tarnow after a raid by a Luftwaffe bomber Gruppe on 5 September 1939. Although some of the bombs have gone wide others have found their mark among the rolling stock. A rising column of dark smoke from burning tank cars is visible from far away.*

Right: *On the same day, another bomber formation attacked the railway sidings at Wreschen. The bomb craters show that this bombing run was flown at an acute angle across the target.*

Top: *As by mid-September 1939 most of the still-serviceable Polish military aircraft had fled to Romania, the advancing German forces captured almost exclusively undamaged training aircraft, such as this Bartel BM 4h.*

Centre: *This propaganda picture carrying the bold caption 'A Messerschmitt diving on a Polish fighter' is relatively easy to unmask: the 'Messerschmitt fighter' must have been a strutted and braced biplane, while its poor victim, a PWS–26 trainer, could never even have aspired to be a fighter!*

Bottom: *After the Polish campaign a number of RWD–8 sports aircraft appeared in Luftwaffe flying training schools. The RWD–8 was used by both Polish flying clubs and the Polish Air Force.*

Top left: Oblt *Wolfgang Falck* as Staffelkapitän *(squadron commander)* of 2./ZG 26. During the period 1940/41 *Falck rendered great service assisting in the build-up of the fledgling Luftwaffe night-fighter force.*

Top right: *Based at Gleiwitz, the crews of 3./ZG 76 flew bomber-escort and low-level attack sorties in the operational area controlled by Luftflotte 4 (General Alexander Löhr). This particular Bf 110B displays the coat of arms of the town of Olmütz, used for a short while as their squadron insignia.*

Bottom: *At the beginning of the Polish campaign a few Do 18Ds of 2./406 coastal-aviation squadron were transferred from List/Sylt to Nest at Jamunder lake from where they flew coastal reconnaissance and search sorties over the Baltic Sea.*

Top left: *An aerial photograph of the Lemberg airfield taken on 8 September 1939 by a crew of 3.(F)/122 long-range reconnaissance squadron. Numerous bomb hits have made the airfield unserviceable, while others have damaged the hangars and accommodation buildings. The drainage system along the southern edge of the airfield shows up as a geometrical pattern.*

Top right: *It is 25 September 1939 and the end is in sight. A He 46 short-range reconnaissance aircraft overflies the badly damaged Polish capital, while some armoured vehicles are advancing into the town through the ruins.*

Bottom: *A Do 17P reconnaissance aircraft shortly before taking off on a sortie over Eastern Poland. On 17 September 1939 Soviet troops crossed the border and incorporated Eastern Poland as part of their territory.*

Top: *Urgent supplies for the flying formations deployed along the eastern border of Germany were carried by transport aircraft. An 'Auntie Ju' (Ju 52/3m) is being loaded up with workshop machinery at Hannover-Langenhagen.*

Centre: *During hostilities in Poland the protection of the Western regions against aerial attacks by British and French aircraft was the task of fighter formations from Luftflotten 2 and 3. The two synchronized-fuselage MG 17s of a Bf 109E of I/JG 53.*

Bottom: *Five pilots of 2./JG 53 at Darmstadt-Griesheim during their tour of duty as 'border guards'. From left: Sepp Wurmheller, Kaiser,* Staffelkapitän *Rolf Pingel, Ignaz Prestele, and Hans Kornatz. The I/JG 53 was deployed at Wiesbaden-Erbenheim and Darmstadt-Griesheim and operated over the Saar-Pfalz sector. [Of the above, Sepp Wurmheller was later to make a name for himself as a great 'dog fighter' in the West until his death in action in June 1944, while after some damage in air combat Rolf Pingel was to force-land his new Bf 109F-1 in the UK on 10 July 1941 – the first aircraft of this type to fall into British hands. Tr.]*

5. The 'Phoney War' in the West and Operation 'Weser Exercise'

During the conflict with Poland a number of flying formations were withdrawn from Luftflotten 1 and 4 and ordered back to their home bases in Western Germany. Following the British and French declaration of war on 3 September 1939 the German command had to take into account intrusions by enemy aircraft, and possibly even an offensive, and for those Luftflotten 2 and 3 stationed in that area were too weak.

However, like the Luftwaffe during the first weeks of hostilities, the Western Allied air forces were under orders to avoid overflying enemy home territory, and bombing attacks were limited to warships and supply vessels. In those months the French *Armée de l'Air* was in the process of re-equipment and modernization, but had nevertheless concentrated all its available formations on operational bases in the eastern and north-eastern areas of France immediately after mobilization. During September 1939, both sides were engaged in border surveillance flights, reconnaissance and nocturnal aerial leaflet-dropping sorties.

On the other hand, the RAF had started its war on 4 September 1939 by a series of daring and costly daytime bombing attacks on the port installations at Wilhelmshafen and on the German warships anchored in the river estuaries as well as carrying out constant flying patrols along the coastal approaches.

During this period, headlines were made by an intrusion of (according the British sources) 22 Vickers Wellington bombers into the German Bight on 18 December 1939. The Luftwaffe single-engined and twin-engined fighters set upon the approaching and retreating Wellingtons with gusto, and reported a total of 36 bombers shot down from a formation of 44. Apart from the contradictory figures regarding the strength of the bomber formation in the announcements of the opposing sides it is astonishing that the Wellingtons overflying Wilhelmshafen did not drop a single bomb. [The actual number of Wellingtons involved and the purpose of this 'raid' have never been cleared up. Even allowing for 'double claims', there must have been more than 22 bombers. It also spelled the end of RAF daytime raids. *Tr.*].

Luftwaffe operations against the British naval forces and ports proved just as costly. Thus, an attack by parts of LG 1, KG 26 and KG 30 on a British naval formation off the west coast of Norway on 9 October 1939 resulted in the loss of nine aircraft, including three crews interned in Denmark. Other Luftwaffe sorties later that month were directed against units of the Royal Navy at Scapa Flow, the Firth of Forth, Shetlands and the Orkneys, as well as, increasingly during the following months, against the convoys and individual ships.

A new term was introduced in the Luftwaffe at that time – 'armed reconnaissance', meaning long-range flights by Ju 88A and He 111 bombers over the western areas of the North Sea as far as the Shetlands. Once a target was spotted, the reconnaissance changed into a bombing attack within minutes.

At the time of this *Sitzkrieg* (lit. 'Sitting War') the tasks of Luftlotte 3 in the West comprised surveillance of the front, reconnaissance flights over France, 'paper flights' (i.e. aerial leaflet-dropping sorties) and countering enemy intrusions. There were almost daily clashes with French fighters near the border areas, while activities on the ground were limited to encounters between opposing reconnaissance and assault groups, without any notable changes in the border position.

It would seem that the almost petty manner in which the occasions when enemy aircraft had overflown neutral borders, such as Luxembourg, Belgium, Holland or Denmark, were given the 'big treatment' in the German press was the result of a deliberate policy to justify Germany's own incursions over neutral territory from time to time.

This kind of aerial warfare dragged on

during the winter months of 1939/40 until the beginning of April 1940, without a single bomb falling in the interior of the three opposing nations. The only exceptions could be considered German bombing raids on the British airfields at Stromness and Kirkwall, close to the Bay of Scapa Flow.

After the operations of the opposing air forces had been effectively limited to attacks on naval targets and ports, and the start of the Soviet-Finnish 'Winter War' on 30 November 1939, both Britain and Germany became concerned to secure the open flank of the Scandinavian area. In the event, the German command started its Operation *'Weserübung'* ('Weser Exercise'), the occupation of Denmark Norway – known according to contemporary style as 'putting them under the protection of the German State' – just a few hours ahead of the British measures. For the first time a branch of armed services that had never before made an appearance on operations was to play the predominant role – transport aviation.

The speedy occupation of both northern countries could not have been carried out without the Kriegsmarine (German Navy) and the KG.z.b.V. (*Kampfgeschwader zur besonderen Verwendung* – Bomber *Geschwader* for special use), the official designation of Luftwaffe air-transport formations. The task of subduing the Danish and Norwegian Air Forces, consisting of only a few squadrons, was entrusted to just one larger Luftwaffe formation, X *Fliegerkorps*, commanded by *Gen Ltn* Hans Geisler. Its subordinated units had to combat both strategic and tactical targets as well as the Allied troops already ashore in Norway and the enemy shipping in Scandinavian waters.

To master the immense transport task the sole existing transport *Geschwader* (KG.z.b.V.1) was augmented by another seven transport *Gruppen* specially formed for this operation (KGr.z.b.V.101 – 107). Equipped with Ju 52/3m aircraft, these were made up from the equipment and crews provided by Luftwaffe multi-engined and blind-flying schools. Another transport *Gruppe*, KGr.z.b.V.108, was equipped with He 59, Do 24, Do 26 and Ju 52/3mW seaplanes.

Loaded with paratroopers, the transport *Gruppen* took off from airfields in the Schleswig-Holstein area during the morning of 9 April 1940. The task of the airborne troops was to drop over the airfields at Aalborg in Denmark and Oslo-Fornebu and Stavanger-Sola in Norway, take them over and secure them for the immediately following transport aircraft with infantry units. Air cover over the airfields and assistance in ground fighting was to be provided by the Bf 110 heavy fighters of I/ZG 76.

Despite some improvisations due to weather conditions all three vital supply bases were taken with minimal losses.

The Norwegian fortifications built into the rocky walls of the Oslo fjord were strongly defended however, and required the assistance of Stukas of I/St.G 1 before the entrance into the fjord was made safe that evening.

The British naval forces operating in the Norwegian coastal waters were a constant threat as, on the one hand, they could land another expeditionary corps and, on the other, they could seriously disturb German operations by naval gunfire and attacks by carrier-borne aircraft. During the following days, these British naval units were attacked by the He 111s of KG 26 and Ju 88As of KG 30, with mixed results.

Despite that, the British managed to land strong forces near Namsos on 15 April and further south in the Romsdalsfjord on 18 April 1940, including some French battalions. At the same time smaller formations of Wellington and Hudson bombers attacked the ports of Trondheim, Bergen and Stavanger, supported by British naval gunfire aimed at the port installations and local airfields. As a result, it was not until early in May 1940 that the British bridgeheads could be cleared and the predominance of their naval formations off the Norwegian coasts finally broken.

While the occupation of the inner regions of Norway against local resistance, occasionally fierce, proceeded according to plan, fighting against the British expeditionary corps supported by heavy naval units in the Narvik area continued until 10 June 1940, already overtaken by German successes in the Western campaign, which had opened a month ago.

The German paratroops, mountain troops and flying formations wrote the final chapter of Operation 'Weser Exercise'. They made it possible to establish and consolidate a series of airfields in the far north which later served as Luftwaffe bases for the surveillance of and attacks on enemy shipping movements in the Eastern Atlantic, in the Polar Sea right up to Spitzbergen, and in the Barents Sea. At the same time southern Norway offered ideal forward bases to Luftwaffe bomber units for raids on Scotland and Northumberland.

Top: *After the air
battle over the
German Bight on
18 December 1939
Obstl Carl
Schumacher
assesses the
performance of his
subordinated units
on the evidence of
crash sites of shot-
down British
Wellington
bombers.*

Centre: *The fighter
squadrons deployed
along the North Sea
coastline had to be
in constant
readiness right
through the winter
of 1939/40. To this
end the first task
every morning was
to clear the
overnight snow
from the aircraft
and the runways, as
here at Westerland.*

Bottom: *Regular
inspection flights to
the operational
bases of individual
Gruppen gave
Luftwaffe staff a
clear picture of local
conditions and the
actual strength of
their formations.
One of the liaison
aircraft used was
this Fieseler Fi 156
GM + AI, one of the
first experimental
examples of the
Storch in Luftwaffe
service, seen here at
Nordholz.*

Top: Major *Fritz Doench (right), together with* Oblt *Philipps and* Oblt *Magnussen, gives a detailed report after the successful attack by his I/KG 30 on the British Home Fleet in Scapa Flow on 16 March 1940. During this phase of the war Luftwaffe activities were limited to attacks on shipping targets in British coastal waters.*

Centre: *The commanding officer of KG 30,* Obstl *Walter Loebel, awaiting the daily operational orders from his Luftflotte headquarters. The KG 30, known as 'Adlergeschwader' because of its eagle insignia, was the first Luftwaffe formation to be equipped with the new Ju 88A dive and horizontal bomber and, together with KG 26 (equipped with He 111 bombers), was considered a special anti-shipping unit.*

Bottom: *The He 60 K6+PH of 1./406 coastal aviation squadron on a patrol flight along the chain of East Frisian isles. On these flights the crews had to keep a special lookout for airmen in distress at sea.*

77

Top left: *From autumn 1939 2.(F)/122 flew long-range reconnaissance sorties with He 111s over northern France. Here F6+CK of Lt Rudolf Hey with its crew, (from left) Fw Lange-Gläscher (pilot), Fw Kiess (flight mechanic) and Uffz Giesa (radio operator/gunner). The observer, Lt Hey, is missing: after all, somebody had to take the picture!*

Top right: *Originally used only as a catapult-launched aircraft on larger units of the German Navy, from early 1940 onwards the Ar 196A was also delivered to coastal aviation groups, where it was intended to replace the stout He 60 biplanes.*

Bottom: *Border patrol over the Bergzabern-Pirmassens area by Bf 109E–3 of 4./JG 53. The 'white 5' is piloted by Uffz Stefan Litjens. In winter 1939/40 the II/JG 53 was based at Mannheim-Sandhofen.*

Top left: *Used for specific point-target defence, 70m³ and 200m³ (2472 and 7063 cu. ft) barrage balloons could operate to a height of 2400m (7870ft). In early spring and summer 1940 a barrage of 200m³ balloons protected the lock installations of Brunsbüttel.*

Top right: *During the period of the 'phoney war' occasional French observation balloons would appear along the Maginot Line – a godsend for the patrolling Luftwaffe fighters.*

Bottom: *In March 1940 Gen Lt Wolfram Freiherr von Richthofen (commanding general of VIII Fliegerkorps) and General der Flieger Albert Kesselring (commander of Luftflotte 2) visited Luftwaffe formations deployed on forward airfields in the Hunsrück-Eifel area.*

Top: *One of the airfields used by Luftwaffe transport* Gruppen *to fly their cargoes to Denmark and Norway was Uetersen. On 7 and 8 April 1940 Ju 52/3m transport aircraft from all over Germany were flying in to land on this airfield in Schleswig-Holstein.*

Centre: *In the morning of 9 April German infantry are loaded aboard the corrugated tin aerial transports, destination Aalborg in Denmark. The largest part of these machines came from twin-engined and blind-flying training schools.*

Bottom left: *During the Norwegian campaign the crews of naval aviation and transport aircraft were still wearing the uncomfortable and cumbersome Kapok life-saving jackets.*

Bottom right: *The modified first prototype of the four-engined Blohm & Voss Ha 139 could be used for eventual special flights and was kept in readiness aboard the motor vessel* Friesenland *in Bremen.*

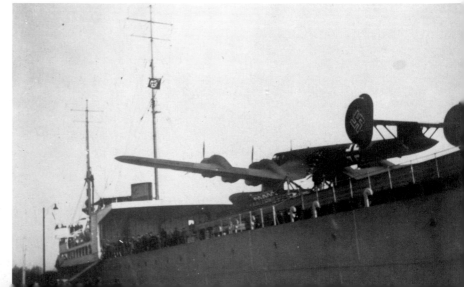

Top: *On 9 April, paratroops were dropped to take the Stavanger-Sola airfield. They had to overcome the resistance of the Norwegian troops guarding it and secure the airfield installations for the subsequent landing of transport aircraft with infantry units. This picture shows the discarded parachutes before their collection that afternoon.*

Bottom: *Ju 52/3m ambulance aircraft stood ready at Uetersen to ferry the wounded to hospitals in Germany. Their intended use was unmistakably signalled to friend and foe alike by their white paintwork and red crosses on wings and fuselage.*

Top left: *One of the formations participating in 'Weserübung' (Operation Weser Exercise, the covername for the invasion of Denmark and Norway) was KG 26, with its Gruppen temporarily deployed at Lübeck-Blankensee, Marx and Jever. Here a He 111H of 9./KG 26 on a compass-swinging base at Jever.*

Top right: *The face of a Ju 88A, an aircraft that first revealed its qualities when flown by KG 30 crews during Operation 'Weserübung'. The Ju 88A could be used as a horizontal and dive bomber and carry a bomb load of up to 2500kg (5512lb).*

Bottom: *This transformer station which reportedly supplied the British-operated propaganda radio transmitter at Tromsö with electric power, was destroyed by German bombers in a low-level attack on 20 April.*

Top: *On 5 May 1940 Allied shipping at Narvik was attacked by He 111s of KGr 100. To be nearer its operational area the whole* Gruppe *had transferred from Aalborg to Trondheim-Vaernes in Norway on 2 May. Here 6N+OH is seen against the background of snowed-in fjords. Note the 'Viking ship' insignia of this specialized pathfinder formation, the only one of its kind in the Luftwaffe at that time.*

Centre: *On 22 April, when it seemed that Narvik would be lost to the Allies, the Do 24V–1 TJ+HR was ordered to fly in two tons of explosives for the destruction of military installations. On this important mission the first Do 24 prototype was flown by Adolf Mlodoch (1st pilot) and Ernst Jurdzik (2nd pilot), who managed to reach their destination and land safely in Beis Fjord despite bad weather conditions.*

Bottom: *A well-deserved Knight's Cross (of the Iron Cross) is awarded to* Lt Werner Baumbach of KG 30. *This young officer excelled himself by daring and successful attacks with his Ju 88A on Allied shipping off the Norwegian coast. A highly decorated* Oberst *by 1944, Baumbach ended the war in command of the unique and mysterious KG 200.*

Top: *The Norwegian Air Force was not backed by an indigenous aviation industry but had to rely on foreign-built aircraft. Here is an abandoned Fokker C VD biplane, formerly part of the obsolete aircraft park of a Norwegian reconnaissance squadron.*

Centre: *A number of Caproni 310 light bombers and touring aircraft were imported from Italy. This civil-registered Ca 310 was captured by German troops at Stavanger fully tanked and ready to take off for England.*

Bottom: *The port and airfield installations at Stavanger were favoured targets for ships of the Royal Navy cruising off the Norwegian shores and British bombers based in the UK. On 17 April British naval gunfire destroyed several He 59 floatplanes and set fire to warehouses.*

Top: *Dievenow in Pomerania was the base of 3./506 coastal aviation squadron which operated over the Baltic Sea right up to the Swedish borders. Here a He 59B–2 floatplane of this unit being lifted by a crane with its engines running.*

Centre: *The Do 18D flying boat could be launched by catapult and was used on long-range reconnaissance tasks. In early spring 1940 aircraft of this type were operational on shipping patrols over the North Sea from their base at List on Sylt.*

Bottom: *After the Western Campaign Do 17s of 3.(F)/22 were transferred for convoy reconnaissance duties to Vaernes and Stavanger-Sola. Shown here are Fw* Müller, *Oblt* Auswäger *and* Uffz Vollmuth *(who took the picture) during a reconnaissance flight with their Do 17Z 4N+AL.*

Top: *A group of*
Luftwaffe fighter pilots
of II/JG 77 taking
things easy in the sun
outside their command
post at Aalborg, the
walls of which are
decorated with three
RAF roundels taken
from shot-down aircraft.
This Gruppe,
commanded by Major
Hentschel, *was*
responsible for parrying
Allied air raids
overflying the northern
tip of Denmark and the
Skagerak.

Centre: *II/JG 77 had its*
first 'red letter day' on
13 April 1940 when its
pilots claimed 15 of 23
Bristol Blenheim
bombers attacking
Aalborg. The remains
of a shot-down
Blenheim IV are being
transported to a
collecting centre the day
after the encounter.

Bottom: *On another*
occasion in April 1940 a
Vickers Wellington
bomber was hit while
attempting to attack the
airfield at Aggersund in
Denmark and had to
make a forced landing
nearby. A Ju 52/3m was
soon on hand to
transport the British
aircrew to a POW camp
in Germany.

6. The campaign in France and the aerial war against England

With the occupation of Norwegian territory making good progress and the securing of the rugged coastline against attacks from the sea, sufficiently strong forces of all three branches of the armed services were at the disposal of the German command for 'Fall Gelb' ('Case Yellow'), the offensive in the West, already postponed several times. In the case of the Luftwaffe, these forces comprised Luftflotte 2 (*General der Flieger* Kesselring) and Luftflotte 3 (*Gen d Flieger* Sperrle), subordinated formations of which had been deployed on operational bases near the Western borders (such as Bassenheim, Ippesheim, Hoppstädten, Mendig, Wengelohr, Bönninghardt and others) since September/October 1939 and had flown border surveillance sorties since that time.

Overnight, on 10 May 1940, the general 'feeling around' by the French, British and German squadrons became an exchange of hefty blows. The element of surprise was completely on the German side in the air as well as with the German army units advancing across neutral Belgium and Holland according to the 'sickle-cut' strategy.

The first blow from the air was struck by air landing troops of *Sturmabteilung Koch* from Fj.Rgt 1 (1 Parachute Rgt) and *Sonderstaffel Schwilden*. The assault detachment led by *Hptm* Koch had to take the blocking fortress of Eben Emael and three bridges across the Albert Canal in Belgium, arriving on the scene in cargo gliders – the first time gliders were used in combat; the paratroops of 7 *Fliegerdivision* had to jump on selected airfields, Rotterdam-Waalhaven, Schiphol and Ypenburg in 'Fortress Holland', take them by assault, and then keep them open for the following transport aircraft bringing in air-landing infantry troops.

The *Sonderstaffel Schwilden* had an even more daring task: with 12 rattling He 59 floatplanes carrying an infantry company it had to land in the middle of Rotterdam, near the two bridges over the Nieuwe Maas and, in the event that it proved impossible to take off again, their crews had to stay and fight together with the infantry holding the bridges until the arrival of reinforcements.

All three operations were surprisingly successful and did much to make the Dutch and Belgian commands feel insecure.

The Luftwaffe bomber force comprising KG 1, LG 1, KG 2, KG 3, KG 4, KG 27, KG 30, KG 51, KG 53, KG 54, KG 55, KGr. 126, KG 76 and KG 77 was in action on level bombing raids all day long on 10 May 1940, at first meeting fierce resistance. At the same time the Stukas of St.G 2, 76, 77, IV/LG 1 and I/186 attacked point targets ahead of the advancing army formations. The fighter escort was provided by JG 1, 2, 3, 26, 51, 53, 54, I/LG 2. II/186 and JGr.21, which, together with the '*Zerstörer*' ('destroyers' or heavy fighters) of ZG 1, 2, 52 and 76 were also tasked with clearing the air space in general. The only ground support/attack *Gruppe*, II (Schl)/LG 2 equipped with Hs 123 biplanes, was on immediate call, while long-range and battlefield reconnaissance was in the hands of the individual *Staffeln* of *Aufkl. Gruppen* 11, 121, 122 and 123. In addition, the short-range (H)-*Staffeln* of *Aufkl. Gruppen* 10, 11, 12, 13, 21, 22, 23 and 41 were directly subordinated to the higher Army formations.

During the early phase of the Western campaign by far the highest loss rate was suffered by KG.z.b.V.1 and the stand-by KGr.z.b.V.9, 11 and 12. These air-landing units were tasked to targets in enemy hinterland and consequently exposed to fierce anti-aircraft and artillery fire. With only about 180 operational aircraft in the Belgian *Aéronautique Militaire* and 130 in the Dutch Air Force, opposition in the air could not amount to much.

On the other hand, the 700 fighters of the French *Armée de l'Air* together with the ten RAF fighter squadrons of the British Expeditionary Force made it very hard going for the Luftwaffe formations during the first few days. Despite that, the Luftwaffe was able to hold its losses within tolerable limits, especially since all Allied airfields in the French–Belgian border area were kept under a hail of German bombs from 11 May onwards.

The obsolete French Amiot, Bloch and Farman bombers had as little chance to break through the Flak defences of the advancing German troop formations as did the Blenheims and Battles of the British Advanced Air Striking Force. More successful were the modern French aircraft, such as the Bloch 151, Dewoitine 520 and Morane-Saulnier 406 fighters, and especially the twin-engined Breguet 693 and Potez 631 multi-purpose aircraft which carried out almost suicidal low-level attacks on river crossings, German troop assembly areas and supply routes. On the other hand, very little was seen of the Dutch and French naval aviation; their aircraft were only partly suitable for operations against land targets.

The Allied air forces in Belgium and France, already considerably decimated by the constant bombardment of their ground installations, received the final death blow during a concentrated Luftwaffe attack on airfields in the Paris area on 3 June 1940 ('Operation Paula'). Following this attack by the two Luftflotten, which was also aimed at the French armaments industry, the French Air Force ceased to exist, despite the fact that according to French data it had achieved 350 certain and 150 probable aerial victories over German aircraft by that date.

During the final days of May, the German forces unexpectedly faced a serious crisis. In order to regroup the Panzerdivisions for the planned attack into central France, Hitler – who was controlling the operations from his HQ 'Felsennest' ('Rocky refuge') at Münstereifel – ordered them to halt on 24 May 1940 and assigned the task of overcoming the British Expeditionary Force surrounded at Dunkirk to the Luftwaffe. It was during these uninterrupted attacks on the Allied troops in this 'cauldron' and their evacuation (Operation Dynamo) which began on 27 May that the Luftwaffe bombers and Stukas met their first Spitfire squadrons. Taking off from bases in southern England, these aircraft provided fighter cover over the port of Dunkirk and caused heavy losses to the approaching German bombers. A few days of bad weather, when no operational flying was possible, also favoured the evacuation, and by 4 June 1940 nearly 340,000 British and French soldiers had been safely taken to the British Isles. As a result of the temporary British aerial superiority over the evacuation zone the operational strength of some German bomber and Stuka formations had shrunk to about half their establishment – a blood-letting that could have had catastrophic consequences.

However, the gradually slackening enemy aerial activity during the second phase of the Western campaign gave the Luftwaffe a breather and an opportunity to replenish its considerably reduced formations. Although target priorities were still concentrated on the strategic side, with airfields and ports all the way down to Marseilles high on the list, from early June onwards the Luftwaffe was being increasingly involved in the tactical support of Army formations. Since the beginning considered as a possible task for the Luftwaffe, in the course of hostilities the use of bomber *Geschwader* as 'flying artillery' would become the rule and push the idea of conducting strategic air warfare, as the Western Allies were clearly doing from 1943 onwards, more and more into the background.

From 14 June onwards Luftwaffe operations were directed in support of the First German Army attacking to break through the Maginot Line from the Saar-Pfalz area, the bombing targets being the fortification belt and its hinterland. The ground attack began on 15 June with an advance into Alsace on a broad front, likewise after a Luftwaffe bombardment.

After the occupation of the French capital on 14 June 1940, and the complete collapse of the French Northern Front, there were increasing signs of disintegration among the retreating, demoralized French units, whose will to resist was to be completely broken not least by the continuous Stuka attacks on the main routes of retreat. Blown up or wrecked aircraft on the occupied French airfields bore witness to the hopeless situation of the *Armée de l'Air* after the collapse of the fronts.

Just as during the Polish campaign, the light and heavy Flak batteries were being used in the forefront of ground fighting from early June, achieving excellent results in direct fire against enemy tanks and bunkers. At this time a number of heavy Flak batteries were withdrawn from the areas of several Armies and deployed as anti-aircraft defence along the Dutch coast, the favourite penetration sector of British aircraft after the RAF Bomber Command had considerably increased the frequency of its nocturnal nuisance raids on German territory. After the first direct British–German confrontation during the occupation of Norway it was no surprise to note the dividing line between military and civilian targets becoming more and more blurred.

While bombing military installations it could happen that bombs would also hit nearby residential areas, the haphazard, if still insignificant, RAF bombing raids on places in Germany in June 1940 which had no military importance whatsoever must be considered as the first sign of an intensification of bombing war against civilian targets.

The almost unhindered advance of the German Army behind constant preparatory attacks by the Luftwaffe far away across the Loire continued until the conclusion of the Armistice on 22 June 1940. By that time the only occasional casualties in the air were caused by anti-aircraft guns. Hostilities in France came to an end on 24 June 1940.

Understandably, operations against England by Luftflotte 2 and X *Fliegerkorps* were far less frequent during the fighting in Scandinavia and the campaign in France. However, with the completion of both campaigns and the acquisition of favourably placed new forward landing grounds in Northern France, Belgium and Holland, aerial activities against the British Isles flared up again and were planned to reach their culmination at the beginning of August 1940. A major effort to prepare all available airfields in these areas for longer-term billeting was initiated after the French capitulation, and a whole series of new airfields laid out with the assistance of the RAD*, the OT** and private contractors.

After a short pause, during which the *Geschwader* and *Gruppen* replenished their personnel- and aircraft-strengths, the formations subordinated to Luftflotte 2 (*Generalfeldmarschall* Kesselring) completed the move into their new bases by 5 August 1940. The final conference of the *Geschwader*- and *Gruppen* commanders followed by a map exercise outlining the attack on the British Isles had been held at the 'Residenz-Theater' in Brussels the day before.

Similar preparations for Operation 'Adler' (Eagle) were made by GFM Sperrle, commander of Luftflotte 3, at his headquarters in Paris. His formations had moved into their bases in the area between Brest and Le Havre.

The newly assembled Luftflotte 5 in Norway

(*Generaloberst* Stumpff) came only partially under consideration for the planned aerial offensive against England: the bomber formations based in Norway would have to fly unescorted there and back because the Bf 109 just did not have the necessary range. Attempts to provide an escort by using the twin-engined Bf 110D long-range fighters fitted with the so-called '*Dackelbauch*' (Dachshund belly) additional fuel tank were abandoned after only a few experimental sorties: the Bf 110D was far too unwieldly for an escort fighter.

The date of 10 August 1940, repeatedly and continually named as '*Adlertag*', the start of the German aerial offensive, is misleading because on that day the Luftwaffe aerial activities were rather less frequent than on previous days. On 11 August however the intensity of German bombing raids on southern England increased noticeably. Operation '*Adler*', to achieve aerial supremacy over England, really began on 13 August 1940 with regard to Operation '*Seelöwe*' (Sealion), the planned invasion of southern England. The Luftwaffe targets on that day were port installations, docks and above all airfields, but the large bomber formations with corresponding fighter escort were foiled by bad weather conditions and the intended 'big blow' did not achieve its expected effect. For instance, Operation '*Lichtmeer*' (Sea of lights), a planned concentrated attack on British airfields northeast of London on 16 August 1940, petered out completely and had to be repeated three days later, with notably better results.

Co-ordination of fighter escorts with the attacking bomber and Stuka *Gruppen* was the responsibility of the so-called *Jagdfliegerführer* or *Jafü* (fighter leaders) appointed in both Luftflotten. On account of the limited flight endurance of the Bf 109 it was vitally important to fix the times of take-off and reaching of covering position literally to the minute to avoid the danger of having 'empty' fighters in the air at the time of contact with enemy fighter defences. The Bf 110 long-range escort fighter proved a failure in that role early on and had to be withdrawn from escort tasks soon afterwards: its performance (speed, manoeuvrability and climb) were decidedly below par to hold its own against the Spitfires and Hurricanes. In fact, the highly vaunted Bf 110 would have required its own fighter escort!

Proportionally the heaviest losses were suffered by the Stuka formations (Stuka-*Geschwader* 1, 2, 3, 77 and IV (St)/LG 1) sent to attack shipping and land targets. The slow Stuka was not only an easy prey to RAF fighters during the approach and departure

---

Reichs Arbeitsdienst, State Labour Service, compulsory to all young men from the age of 17 onwards. *Tr.*
**Organisation Todt*, a State-wide organization responsible for all kinds of strategic and military construction works from the Autobahns via the 'West Wall' ('Siegfried Line') to the 'Atlantic Wall'. Its founder and leader, Dr Fritz Todt, was killed in an air crash in February 1942, after which OT was taken over by the Ministry of Munitions and Albert Speer, but keeping its original name and OT armband till the end. *Tr.*

flights but even more so in a dive when the Ju 87 pilot could not make the slightest evasive movement and thus did not have a chance against the pursuing and much faster RAF fighters.

The operational diary of the three Luftflotten notes a total of 4787 aircraft sorties in August 1940:

404 in concentrated large-scale attacks
3898 against 'destruct' or secondary targets
239 against shipping targets
246 minelaying sorties

The approach and departure flights over the sea made it essential to have an efficient air-sea rescue service as quickly as possible to save the crash-landed or baled-out aircrews. In the case of Luftflotten 2 and 3 this task was carried out by the flying boats and floatplanes of *Seenotdienstführer 3* (air-sea rescue leader) based at Cherbourg.

The mining of British ports and river estuaries was started on a larger scale by 9.*Fliegerdivision* late in July, reaching its peak in November 1940 with 605 aircraft sorties – nearly all of them in bad weather or at night.

During the month of August the aim of the Luftwaffe command was to hold down and destroy the defensive power of the RAF Fighter Command. Accordingly, airfields and aircraft and aero-engine production facilities were the priority targets of Luftwaffe formations large and small. This changed late in August when a series of RAF night bombing raids on Berlin (beginning 25 August 1940) induced Hitler to alter the targets of the 'England offensive' and instead of combatting the installations hit back with bombing raids on London. The word 'revenge' was spoken publicly for the first time since the beginning of the war and was destined considerably to influence the further course of the air war. It happened at a time when the fighting potential of RAF Fighter Command was already so exhausted that it would have needed only a few more well-concentrated and escorted Luftwaffe operations to gain absolute aerial superiority over the southern sector of the British Isles and provide positive air cover for the planned landing operation. But it was not to be.

If until then it had been the occasional miss that had hit dwelling houses near some military installations, the Luftwaffe now began to carry out planned bombing raids on '*Loge*', the Luftwaffe target covername for London. A relatively weak attack on 5 September was followed by the first large-scale bombing raid on London during the afternoon hours of 7 September. Carried out under the key-word of '*Vergeltung*' (revenge), this attack was flown by parts of eight Luftwaffe bomber *Geschwader* escorted by fighters from JG 3, 27, 51, 52, 53, 54 and 77 as well as ZG 2. Then followed a period of bad weather which prevented further daily large-scale bombing raids on London until 19 September; the bombing could only be carried out by formations of *Staffel* strength.

Nevertheless, the largest number of operational Luftwaffe bombers of the entire Operation '*Adler*' was recorded on 15 September. On that day the British reported the loss of 27 of their own fighters as against 56 German bombers shot down; the German reports spoke of 79 British aircraft as against just 43 German losses. According to the official returns of the Quartermaster General of the Luftwaffe, the actual total German losses that day were 49 aircraft, i.e. 15% less than British claims. At the same time the Luftwaffe claims of 79 British aircraft must also have been considerably exaggerated – the propaganda people had a finger in the pie on both sides.

From 20 September 1940 onwards the German air raids on England were shifted to late evening and night hours. This change required the assistance of the 'pathfinders' and 'illuminators' of KGr.100 which used the 'X' and 'Y' navigation methods to indicate target areas with flares and ground markers for the bombers following after them.

In daytime, the aerial offensive against England was left almost exclusively to *Jabos* (from *Jagdbomber*, fighter-bombers), single-engined Bf 109s fitted with bomb racks together with bomb-carrying Bf 110s of *Erpr. Gruppe* 210 (trials unit) commanded by *Hptm* Lutz. The only other 'daytime operators' were the leading crews of some *Geschwader* who would fly alone with their He 111 and Ju 88 bombers in bad weather conditions, attacking such vital industrial targets as the Standard Motor Works at Coventry, Supermarine at Southampton-Eastleigh, Rolls-Royce at Derby, Percival at Luton or Leyland Motors at Leyland, to mention only a few, bombing them in very risky low-level attacks. The names of *Hptm* Storp, *Oblt* Rinck, *Hptm* Dürbeck, *Oblt* von Buttlar, *Ofw* Kless and *Oblt* Elle with their crews are mentioned here as representative of the many other highly experienced airmen who participated in these operations.

The total of 7620 aircraft sorties in September, 9911 in October, 6778 in November and 3884 in December 1940 are still quite high compared to 2468 in January and only 1401 in February 1941, signalling a retrogressive tendency in the success expectations of the German leadership.

By this time formations of X *Fliegerkorps* had already been moved to the Mediterranean zone to render assistance to the Italian allies who had got into a serious mess in North Africa. Because of difficulties in communications and radio control the Italian effort to join the aerial offensive against England was found to be more of a hindrance than real help. After Italy had joined the war on 11 June 1940 formations of their *Regia Aeronautica* had become active, raiding the islands of Corsica and Malta, and also British positions in Egypt and on the Dodecanese islands with their three-engined Savoia-Marchetti bombers. In October 1940 the C.A.I. (*Corpo Aero Italiano*) arrived in Belgium as an Italian contribution to strengthen Luftflotte 2. This Corpo comprised the *Stormi* 13 and 43 equipped with Fiat BR.20M bombers, *Gruppo* 18 with Fiat CR.42 biplane fighters and *Gruppo* 20 with Fiat G.50 monoplane fighters – which were forthwith redesignated KG 13, KG 43, 18./JG 56 and 20./JG 56 and incorporated into Luftflotte 2. At the same time the *Squadriglia Ricognizione* with its five Cant Z.1007s became *Aufklärungsstaffel* 172.

While participating in nocturnal Luftwaffe raids on England on 24 and 29 October, 11 and 23 November and 21 December 1940 the Italian airmen, not familiar with weather conditions in Northern Europe, soon came to realize that this kind of aerial warfare was no child's play. By Christmas 1940 all their units except 20./JG 56 had returned to the warmer Mediterranean climes, where the military situation had changed for worse in the meantime.

The first attempts at long-range night fighting were made late in September 1940 by some Ju 88Cs, Do 17Z–10s, and Do 215s of I/NJG 2 commanded by *Hptm* Hülshoff. Crewed by airmen with first-rate blind flying experience, these aircraft would take off from their base at Gilze-Rijen in Holland and set course towards known RAF bomber bases, being guided by intercepted radio signals emitted while tuning the RAF bomber radio sets on the ground, the idea being to stalk and shoot down the bombers while they were taking off and assembling near their bases. The most widely used tactic was however to 'hang on' behind an RAF bomber leaving German territory and attack it in textbook style as it was approaching its base to land, with all navigational lights and airfield lighting switched on. [These were the first long-range 'night intruders', a completely new kind of aerial warfare which could have been very successful if Hitler had not stopped it: 'German people wanted to see the aircraft being shot down.' A few years later, RAF Beaufighter and Mosquito long-range night intruders became the scourge of German night fighters and all other kind of night flying operations. *Tr.*]

At that time there were no useful night interception devices in Luftwaffe service; the early night interception radar sets were still in development and the so-called '*Spanner-Anlage*' ('Gaffle' or 'Trigger') infra-red detection device was still under test.

Photographic reconnaissance and target-effect pictures were an essential part of this aerial offensive, and aircraft of the long-range reconnaissance *Staffeln* were operating almost daily, flying alone and without fighter escort, all over the British Isles to take the necessary photographs, with correspondingly heavy losses.

The short-range (H)-*Staffeln*, still equipped with the Hs 126 high-wing monoplanes, were confined to flying local reconnaissance sorties over the Channel together with coastal aviation units and camouflage-checking flights over own military installations.

That Operation '*Seelöwe*' (Hitler's War Directive No. 16 of 17 July 1940) was not carried out was in part because the Luftwaffe did not succeed in gaining aerial mastery over southern England and in part because of doubts the German naval command had regarding the availability of the necessary transport capacity for such an undertaking. With all that, the main reason for this postponement of the intended invasion of England was most probably Hitler's principal goal, war against Communist Russia, which had made its expansionist intentions quite clear by its policy of annexation after the Soviet–Finnish Winter War of 1939/40 and then the occupation of the Baltic States in June 1940.

Top: *The element of surprise created by the assault glider proved itself on the very first day of the Western campaign. The Assault Detachment 'Granit' was landed in several DFS 230A gliders atop the Belgian fort of Eben Emael on 10 May 1940 and succeeded in overcoming the fortress garrison troops within a few hours.*

Bottom left: *An aerial photograph of the blocking fortress of Eben Emael, situated immediately next to the Albert canal. The nine DFS 230A gliders landed silently between the armoured cupolas of this installation; the Belgians had not anticipated an assault from the air.*

Bottom right: Oblt *Rudolf Witzig, leader of the Assault Detachment 'Granit'. The towing cable of his glider was torn shortly after taking off from Köln-Ostheim and Witzig landed on the fortress plateau with a replacement aircraft about three hours later than his comrades.*

Top: *Following a bombing raid by II/KG 4 'General Wever', a battalion of paratroops jumped over the Dutch airfield of Rotterdam-Waalhaven. They were reinforced by an infantry company landed in Ju 52/3m transports while the fighting was still in progress.*

Bottom: *On 10 May 1940 a number of Ju 52/3m transports, hit and practically ripped apart by Dutch trench mortar fire while landing, later had to be dragged aside to clear the runways for the incoming aircraft.*

Top: *A number of Ju 52/3m transports that had landed on the airfield at Ypenburg near Delft were shot in flames by Dutch artillery. As a result, the few surviving German infantry troops were unable to take the airfield as planned.*

Centre: *Heavy Dutch artillery fire also destroyed some Dutch aircraft which had been parked alongside the runways, such as this twin-fuselage Fokker G1 heavy fighter.*

Bottom: *Two He 59 floatplanes of* Sonderstaffel Schwilden *taxied into each other near the Vaste Spoorwegbrug bridge over the Neuve Maas river and had to be abandoned. According to plan, twelve of these twin-engined floatplanes had to land a company of infantry troops in Rotterdam.*

95

Top: *Without the 'Schwarzen Männer', the ground crew and maintenance personnel, the Luftwaffe probably would have had to move on foot. Here* Obgefr Bernhard, *one of the armourers of I/JG 77, checks the mechanism of a wing-mounted MG 17 machine gun of a Bf 109E.*

Centre: *A harvest of medals for 3./JG 54 following a successful low-level attack on the Luxeuil airfield. Back row, from left:* Uffz *Krantz, squadron commander* Oblt *Schmoller-Haldy, Lt Kinzinger, Lt Angeli, Lt Witt. Front row, from left:* Uffz *Windisch and* Fw *Knedler.*

Bottom: Ofw *Grimling of I/JG 53 was killed in air combat on 14 May 1940 and buried next to his crashed Bf 109E fighter. Many such 'missing in action' cases were cleared up by advancing Army units encountering air-crash sites on the way.*

96

Top: *Badly shot up on a reconnaissance flight, a returning Do 17P catches fire on landing back at the base. In this case, the crew managed to get out of their aircraft in time.*

Centre: *Two 50kg (110lb) bombs under the starboard wing of a Hs 123 of II(S)/LG 2 This unit was the only* Gruppe *in the Luftwaffe equipped with these sturdy biplanes – and also the only dedicated ground-support formation at that time. Their baptism of fire had been in Poland. [The inscription on the outer bomb reads 'for 4 o'clock tea!' Tr.]*

Bottom: *Ju 87 Stukas of VIII* Fliegerkorps *dive-bomb a fortification at Namur on 17 May 1940. It was mainly the demoralizing effect of this precision weapon that led to the premature capitulation of such fortifications.*

Top: *Preparations for take-off on another reconnaissance flight to observe the advancing Panzer spearheads on the ground: the Hs 126B T1+KH of 1.(H)/10 'Tannenberg' on 16 June at Choiselle in France. A Fi 156C Storch liaison aircraft is parked nearby.*

Centre: *An obsolete Belgian LA-CAB GR–8 biplane bomber trapped under a collapsed hangar roof at Brussels-Evere, found by advancing German troops. In fact, Belgian operational squadrons were mostly equipped with British aircraft types.*

Bottom: *A number of undamaged Renard R 31 parasol reconnaissance monoplanes were discovered in a hangar at Nivelles. These machines were of a local Belgian design and equal to similar aircraft of other combatant nations.*

Top: *The RAF fighter squadrons sent to the Continent hardly had a chance against the swarms of Messerschmitts. Despite its eight wing machine guns this Hurricane I of the BEF (British Expeditionary Force) got the worst of it in air combat and the pilot had to make a forced landing near Sedan.*

Centre: *Two rows of bomb explosions erupt across the airfield at Reims. According to a pattern practised in Poland, the emphasis also in the West was first to knock out the air bases and landing fields of the enemy air forces.*

Bottom: *The French Morane-Saulnier MS 406 was a first-rate fighter. Early in 1940 two squadrons equipped with these machines were formed of Polish fighter pilots who had managed to reach France and were incorporated into French fighter groups. This particular MS 406 belonged to one such unit (note the red-and-white Polish national insignia on the fuselage).*

Top: *After Churchill had turned down the proposed evacuation of British troops from Calais they came under ever fiercer Luftwaffe attacks. On 26 May Ju 87s of St.G 2 'Immelmann' carried out a series of 'rolling' dive-bombing attacks on the British troops defending the port installations.*

Centre: *The effect of one 500kg (1102lb) Stuka bomb near the railway station of Evreux. Accurate dive bombing by Ju 87s cut many French railway lines interrupting or stopping completely vital supplies reaching the fighting troops at the front.*

Bottom: *As a light fighter, the Caudron C 714 played no role in the French aerial defence. This C 714 was found abandoned at Evreux on 22 June 1940. [One squadron of Polish fighter pilots in France was equipped with the C 714, and the cut-out square piece of doped fabric on the fuselage exactly matches the Polish national insignia – therefore, this is probably an ex-Polish-flown C 714. Tr.]*

Top: *The intended MS 406 replacement was the Arsenal VG 33, but by May 1940 only a few of these new fighters had been completed and none delivered to fighter formations.*

Centre: *The Bf 110 counterpart in the* Armée de l'Air *was the Potez 630. Fitted with heavy fixed armament in the nose, it was also used as a night fighter. [The aircraft shown here is a Potez 63 series demonstration or special 'courier' machine; it carries no armament and is not camouflaged. Tr.]*

Bottom: *The Loire-Nieuport LN 401 was intended to be taken into service by the French naval aviation as a dive bomber but fell short of expectations. The pilot of this particular LN 401 had to make a forced landing near Brest after an encounter with a German fighter.*

[The LN 401 was in service with the French Aéronavale; 82 had been ordered by July 1939. The similar LN 411 was the French Armée de l'Air *variant (less all naval equipment), rejected in October 1939, with all completed machines going to the land-based French naval aviation units.*

Two escadrilles (AB2 and AB4) participated as low-level attackers in ground fighting beginning 19 May 1940, when 10 out of 20 attacking LN 401s were shot down. Tr.]

Top: *An oil-storage depot at Royan, near the Gironde estuary, set on fire by German bombers. Deep penetration raids by the Luftwaffe contributed to the almost chaotic collapse of the French armed forces.*

Centre: *Hidden deep in the German command safes was a plan bearing the covername 'Tannenbaum' ('Fir tree'), a pincer attack on the French through Swiss territory. Bf 110Cs of I/ZG 52 carried out constant surveillance patrols along the Swiss border.*

Bottom: *For their part, fighter patrols of the Swiss Confederation kept an eye on their own airspace. Encounters between their D 3800 (MS–406) fighters and German aircraft which had crossed the border took place on several occasions, resulting in losses on both sides.*

Top: *A scene like this at Bordeaux-Merignac was repeated on many French airfields after the Armistice on 22 June 1940: scores of French aircraft would be gathered in one corner of the airfield awaiting their disposal – mostly scrapping. In this case, the parked machines include a selection of obsolete Farman 222 and Bloch 200 and some more modern Potez 631 and Bloch 131 bombers.*

Centre: *Luftwaffe activities against England began gaining momentum immediately after the end of hostilities in France. Uffz Sellhorn belonged to 1./St.G 77, one of the Stuka squadrons that was badly thinned out during this period. Maltot near Caen, September 1940.*

Bottom: Oblt *Martin Schmid, commander of 8./St.G 51 survived 14 highly dangerous Stuka raids on the British Isles. As part of the Troop Welfare action his unit was visited by a group of showbusiness people in September 1940, including the famous screen actress Olga Tschechowa, seen here getting acquainted with the innards of a Ju 87B Stuka at Norrent-Fontes.*

Top: *The Do 215B–4s of
4.(F)/Aufkl.Gr.Ob.d.L. (see Glossary)
based at Merville carried out
increasingly more frequent flights over
the British Isles: it was important to
complete the target reference maps for
the forthcoming Operation 'Adler'. The
ventral bulge under the forward
fuselage of this Do 215B–4 of the above
unit contains the Rb 30 × 30 automatic
aerial camera.*

Centre: *An atlas of aerial photographs
of all potential enemy countries was
surreptiously compiled by German civil
aircraft belonging to the*
Versuchsstelle für Höhenflüge
*(Experimental High Altitude Flying
Centre) led by Theodor Rowehl before
the war. In 1939 he took over command
of the* Aukl.Grupe Ob.d.L.

Bottom: *The unarmed Junkers Ju 86R
high-altitude photographic
reconnaissance aircraft were out of
reach of the enemy defences, and
operated right up to the Scottish
Highlands. Their service ceiling of
13,000m (42,650ft) was just too high for
the RAF fighters.*

Top: *On 13 August 1940, the day when Operation Adler, the German aerial offensive against England, began, crews of I/KG 76 bombed Redhill airfield. Here* Oblt *Rudolf Hallensleben, squadron commander of 2./KG 76 discusses a point with his ground-crew chief in front of his Do 17Z–2.*

Centre: Major *Erich Kaufmann (right), commander of I/KG 53 'Legion Condor' reporting to* Generalfeldmarschall *Kesselring about operations over England in September 1940, watched by* Oberst *Stahl, commander of KG 53 (centre).*

Bottom: *A He 111P of 1./KG 55 unloads its eight 250kg (551lb) bombs over the Bristol aircraft works at Filton on 25 September 1940. In that month, the Luftwaffe bombers began increasingly to mount raids on the British Isles at night.*

Top: *The new posts of Jafüs (Jagdfliegerführer, Fighter Leaders) were created before the aerial offensive against England to control tactical operations of fighter Gruppen subordinated to each Luftflotte deployed in the West. The operational headquarters of Jafü 2 (Luftflotte 2) was at Le Touquet, south of Boulogne.*

Centre: Generalmajor *Theo Osterkamp, a Marine fighter ace of the First World War, was appointed Jafü 2. In his area operated the JG 3, 26, 51, 52 and 54, as well as ZG 26.*

Bottom: *The fighter career of* Oblt *Helmut Wick was comet-like. A squadron commander of 3./JG 2 'Richthofen' at the beginning of the French campaign, he was promoted commander of the entire JG 2 on 20 October 1940 and reported missing in action over the Channel on 28 November 1940. By that time he had achieved 56 confirmed aerial victories.*

Top: *With the inevitable cigar in his mouth a shirtless Major Adolf Galland (centre) and a group of his officers observe an aerial combat right over their home base at Caffiers, close to Pas de Calais in July 1940. At that time Galland was commander of III/JG 26 'Schlageter'.*

Centre: *The billets and dispersal areas of Luftwaffe flying formations were occasionally marked with quite original indicator boards. In this case it is clearly the domain of I/JG 26 'Schlageter' under its then commander Rolf Pingel. The inscription on the board reads 'Devil's Village, Adm. area Galland, District Pingel, Burgomaster Munde'.*

Bottom: *Galland's personal BG 109E–4 during his time as commander of JG 26 'Schlageter' at Audembert. The rudder displays 60 victory stripes, with dates.*

Top: *Time and again,
Luftwaffe bombers were
launched against supply
shipping off the British
Isles. On 15 July 1940 it
was the turn of this 5000-
ton freighter, sunk by
bombs of two He 111s of
5./KG 27 'Boelcke'.*

Centre: *He 111s of
I/KG 53 'Legion Condor'
in sandbag 'boxes' and
under camouflage netting
at Vendeville in
September 1940. These
measures had become
necessary since the RAF
had started surprise
attacks with light
bombers on Luftwaffe
airfields in northern
France.*

Bottom: *Strengthened
defensive armament on
the Do 17Z–2 of 9./KG 76
in the form of additional
machine guns in beam
windows. All light
coloured parts of this
aircraft, including the
usually sky-blue fuselage
belly, have been covered
by black night-camouflage
paint.*

Top: *A Do 17P of 2.(F)/123 long-range reconnaissance squadron in transfer flight to Jersey for a night reconnaissance sortie over England. It carries four 50kg (110lb) LC 50F parachute flares under the external under-fuselage bomb racks.*

Centre: *Returning from a raid with a shot-up engine, a Ju 88A of I/LG 1 crashes on landing at Orléans-Bricy on 22 outright in the inferno.*

Bottom: *Ju 88A–1s of I/LG 1 are being loaded up with 1000kg (2205lb) SC 1000 bombs carried on external bomb racks. In this case, the night camouflage paint has been used to cover the nationality markings as well.*

109

Top: *The three-engined Italian Cant Z.1007 Alcione (Kingfisher) bombers were not seen very often in France. They would not have done all that well if they had tried to fly reconnaissance sorties over England – the waiting Spitfires and Hurricanes could hardly miss such 'barn-doors'.*

Centre: *It was of course a different matter when it came to wartime propaganda. This is how the noted artist Hans Liska visualised a combined German–Italian attack flown by Fiat BR.20 Cicogna (Stork) and Ju 88A bombers. In reality nothing like it probably ever took place – at the very least the aircraft would have been flying in the same direction!*

Bottom: *The He 59 floatplanes used by the German ASR service in autumn 1940 were unarmed and displayed large red crosses, which did not protect them from being attacked by British fighters on several occasions.*

Top left: *In summer 1940* Reichsmarschall *Hermann Göring himself paid a visit to the Luftwaffe formations stationed along the Channel coast.*

Top right: *Led by its commander Obstl von Chamier-Gliscinski (left) KG 3 flew day and night bombing raids against the British Isles throughout the winter of 1940/41. Von Chamier is seen here talking to General Bruno Loerzer, commanding general of II.Fliegerkorps.*

Bottom left: *The 'Channel trousers' and life-saving jacket became the necessary requisites of all Luftwaffe aircrews participating in operations against the British Isles. The equipment carried included dye satchels, signal flares and a flare pistol, all intended to help the ASR aircraft crews to spot the airmen in distress in the sea.*

Bottom right: *This ditty reminded the 'shooting fighter' fraternity to whom they owed thanks for their success – their wingmen, who had to protect the backs of their No. 1s.*

111

Top:*For Operation 'Seelöwe' (Sealion) the Kriegsmarine gathered all kinds of landing boats, ferry barges and rebuilt motor freighters in Dutch and Belgian ports beginning July 1940. However the planned landing in southern England never materialized.*

Centre: *Despite many propaganda photographs which showed He 111s with this* Geschwader *insignia over England, there was no such unit. This insignia was painted on a number of aircraft participating in the film* Kampfgeschwader Lützow.

Bottom: *By April 1941 a cold wind was already blowing around the ears of Luftwaffe fighter pilots and consistent aerial victories against the revitalized RAF were few and far between. Three of the more successful pilots of JG 26 'Schlageter' (left to right): Hptm Rudolf Bieber, Hptm Rolf Pingel (commander I/JG 26) and Oblt Josef Priller (squadron commander of 1./JG 26).*

Top: *A small number of Bréguet 521 Bizerté flying boats found intact after the French capitulation were a welcome addition to the Luftwaffe ASR squadrons. Four of these large flying boats were in service at the Brest-Poulmic Luftwaffe ASR centre.*

Centre: *The Do 24T, evolved by Dornier for production under licence in the Netherlands before the war, proved itself as a fast, seaworthy and roomy ASR aircraft in service. Fitted with three gun turrets, it also proved difficult for attacking fighters to digest.*

Bottom: *Another odd-looking big bird appeared at 2.(F)/Aufkl.Gr.Ob.d.L. – the Blohm & Voss Ha 142V–2/U, a rebuilt prewar long-range mailplane. It was intended to fly reconnaissance sorties over England but wiser councils prevailed and in the event it was only used for patrol and search flights over the sea and some transport tasks.*

113

Top: *Long-range night-fighting operations began rather modestly with a few Do 17Z aircraft. In addition to their bomb load, they were fitted with a fixed 20mm MG–FF cannon in the bottom part of the fuselage nose glazing as an offensive weapon. So modified, a Do 17Z–2 is being covered with 'midnight black' camouflage paint at Gilze-Rijen in August 1940.*

Centre: *Fitted with the grafted-on solid nose of a Ju 88C heavy fighter, the Do 17Z–2 became Do 17Z–7 Kauz I. These machines were used by I/NJG 2 for long-range night intruder sorties over England.*

Bottom: *After the Bf 110 long-range escort fighter had proved unsuitable in that role it was found useful as a night fighter. This is an 'airman's monument' (Luftwaffe slang), a Bf 110D G9+CB 'erected' by a pilot of III/NJG 1 on 31 March 1941 at Schaffen-Diest in Belgium.*

Top: *A Luftwaffe aerial photograph of the airfield at Flookborough partly covered by scattered clouds. Situated a few kilometres north of Barrow-in-Furness, it had three angled overlapping runways.*

Bottom: *Barrow-in-Furness. Interpretation of this aerial photograph taken on 11 June 1941 revealed the following ships in the port and the docks: an aircraft carrier (1), one light cruiser (2), four destroyers (3, 4), two submarines (5, 6) and several freighters.*

Top: *An aerial photograph of the weapons training centre and firing range at Eastchurch on the east coast of England, south of Southend. Such conspicuous military installations remained the principal targets of the Luftwaffe bombers until early September 1940.*

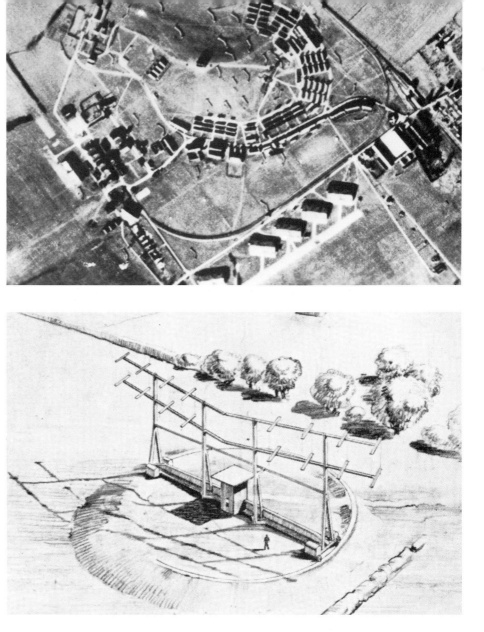

Centre: *A device to lead and control the bombers and pathfinder aircraft was evolved in 1939/40, the FuSAn.721 known as 'Knickebein'. It transmitted a localizer beam across the target which was followed by the attacking aircraft. This drawing depicts the 'Knickebein' beacon on its turntable.*

Bottom: *On returning from a 'Knickebein' – guided night-bombing raid on the Rolls-Royce works at Derby this He 111H piloted by* Ofw Wolter of I/KG 53 *crashed north of its home base of Vitry-en-Artois on 28 August 1940. There were no survivors.*

7. The Balkans, Greece, Crete and the Mediterranean

Optimistic daily reports from the Italian allies could not hide the fact that in the Italian sphere of interest south of the Alps things were not what they were claimed to be. The offensive in Greece, started by the 'Duce' early in 1941 without any coordination with the German government, had got bogged down, as had the Italian attack in North Africa, which was held up at the Libyan–Egyptian border. A *Fliegerkorps* and one Panzer-division, offered by Hitler to the Italian dictator to support the operations on North Africa, were turned down by Mussolini. Now, however, as the Italian divisions were forced to give up one position after another in Greece and North Africa and Italian predominance in the Mediterranean area become uncertain, German assistance to stabilize the situation was most welcome.

To protect and keep the oil output going in Romania, the import of which was of great interest to Germany, the German government had offered the Romanian King Carol the assistance of German army and Luftwaffe missions already in September 1940, and both had been accepted on most friendly terms. The presence of British bases in the Mediterranean area and the unmistakable demands of the Soviets to have a greater influence in Bulgaria, as well as the occupation of the Straits of the Black Sea and the Sea of Marmara would have certainly be taken as a threat to the south-eastern area of the Mediterranean basin and induce the German government to take countermeasures. Operation 'Sonnenblume (Sunflower), military assistance to the Italians, was about to begin.

Luftwaffe participation had already started in December 1940 when the entire III/KG.z.b.V.1 was deployed south to assist the inadequate transport units of the *Regia Aeronautica* in supplying the Italian invasion forces in Greece by carrying out daily transport flights to Tirana and Dewoli in Albania. The following month X *Fliegerkorps* also began to move south, settling on various bases in Italy. The formations involved were III/ZG 26, I/St.G 1, II/St.G 2, II and III/LG 1, II/KG 26 and 1(F)/121, which formed the advance guards of further Luftwaffe units operating in that area.

The first Luftwaffe formation sent to Romania late in November 1940, after it and Hungary had joined the Three Power Pact, was I/JG 28 (III/JG 52) equipped with Bf 109Es, based at Bucarest-Pipera for the protection of the oil fields. Bulgaria and Yugoslavia entered into partnership with the Axis Powers in March 1941 and seemed to strengthen their positions in south-eastern Europe. But then came an anti-government *coup* in Belgrade on 27 March 1941, followed by a change of government and a Soviet-Yugoslav friendship pact, which created an unforeseen, critical situation in the Balkans. To clear it up, Hitler saw only one possibility: the occupation of Yugoslavia, and of Greece, where the British had established bases beginning 7 March 1941. The Luftwaffe forces in that area comprised Luftflotte 4 and VIII *Fliegerkorps*, parts of which had been stationed in Romania and Bulgaria since January 1941 (II and III/JG 27, III/St.G 2, I/St.G 3, II/ZG 26 and I/LG 1). These were now speedily reinforced by other formations to form a powerful air strike force.

On 6 April 1941 Luftwaffe formations overflew the Serbian border – Operation 'Marita' had begun.

Targets for the first day were the Yugoslav airfields and the fortifications of the capital, Belgrade. Most of the enemy aircraft were destroyed by Stuka and low-level attacks at their parking places on airfields, and some Bf 109E fighters, exported to Yugoslavia in 1940, were shot down in air combat. On that first day, the Luftwaffe units reported 32 total losses due to enemy action – the highest loss quota of the entire Balkan campaign. On the following days it was practically only anti-aircraft artillery that the German aircrews had to watch out for; hardly another Yugoslav aircraft was seen in the air.

This changed during the attacks on military installations in Greece beginning 14 April 1941, when the daily Luftwaffe loss rate increased quite considerably. From then on Luftwaffe aircrews had to deal with Greek and British fighters.

Next to road and rail communications the targets of Luftwaffe bomber and Stuka *Gruppen* were also troop assembly areas and transport ships with which the British tried to land reinforcements and material supplies. The Yugoslav army capitulated on 17 April. The Italian contribution in that campaign was limited to the advance by the Italian Second Army along the Dalmation coastline; their main interest lay in territorial claims.

While turning the flanks of Greek defensive positions the Luftwaffe bombers were typically once again used for the tactical support of army units, just as shortly before in Yugoslavia. On the other hand, the operational plans of X *Fliegerkorps* in southern Italy were entirely strategic. During the whole Operation *'Marita'* III/St.G 1, III/KG 30, II/LG 1, III/LG 1, II/KG 4, II/KG 26 and III/KG 3 had, together with Italian squadrons, attacked the British supply base Malta with its intertwined airfields, and thus tied down the RAF fighter squadrons there and prevented their transfer to the Peloponnese peninsula.

The RAF squadrons stationed in Greece, equipped with Gladiator and Hurricane fighters, Blenheim and Wellington bombers and Lysander short-range reconnaissance aircraft, flew daily and very costly attacks against the rapidly advancing German divisions, as a result of which the RAF formations were practically eliminated in a few days. The remaining 30 or so aircraft were withdrawn to Malta shortly before British troops began to evacuate Greece (22 April 1941). The Greek squadrons, equipped with obsolete machines, had no chance against the superior Luftwaffe in any case. On 23 April came the third and final part-capitulation of the Greek armed forces, and once again there was a stable political situation in south-eastern Europe. As two 'foreign craft' anchored side by side there now stood only the two strong British island bases of Malta and Crete – together with a number of smaller islands – right on the Axis doorstep.

As a springboard to the North African theatre of war and the strategically valuable Suez Canal, the island of Crete in the southern Aegean was of great importance to the Luftwaffe. Crete could only be taken by a combined navy and army invasion or in an air-landing operation. In both cases the risk was great: after all, most of the 50,000 or so British troops evacuated from Greece, were now in Crete, and further reinforcements could be expected.

The German command decided in favour of an air landing which, known as Operation *'Merkur'*, would be supported by smaller seaborne landing units transported in requisitioned boats.

The air-landing force came from XI *Fliegerkorps* (formerly 7.*Fliegerdivision*) commanded by *Gen Oberst* Kurt Student which with its subordinated transport *Gruppen* had to carry the paratroops and parts of 5 Mountain Division. As the Greek airfields with concrete runways (Eleusis, Tatoi and Larissa) had been taken over by Luftwaffe bomber formations, the 500-odd available Ju 52/3m transporters had to be distributed to temporarily hardened airfields around Athens. The transport *Gruppen* deployed for this task were I and II/KG.z.b.V.1, I/KG.z.b.V.172, KGr.z.b.V.40, 60, 101, 102, 105 and I/LLG 1 towing *Gruppe* for the DFS 230 cargo/assault gliders.

On the morning of 20 May 1941 the first fully loaded Ju 52/3m transports lifted off from their sandy airfields, pulling behind enormous dusty clouds which completely blinded the pilots of the following *Staffeln* and delayed the assembly after take-off. While the slower airlanding formations were on their way, the Luftwaffe bombers of I and II/KG 2 (Do 17Z), II/KG 26 (He 111H) and the entire St.G 2 concentrated on just one target: the airfield at Maleme and its surrounding area. At the same time the Bf 110s of I and II/ZG 26 together with Bf 109s of II and III/JG 77 worked over known anti-aircraft and artillery positions in low-level attacks.

The first DFS 230 gliders with paratroops of the Assault Regiment landed while the last low-level strikes were still being flown. Their task was to hold down the enemy resistance during the following parachute jump and to clear the airfield for the Ju 52/3m transports carrying the mountain troops and supplies.

German casualties against the grimly fighting British defenders were alarmingly high during the first few days. Most of these losses were incurred during the airborne assault when the paratroopers of 1, 2 and 3 Parachute Regiments were helplessly dangling under their 'chutes, offering an ideal target. After the survivors had formed small combat groups and begun to clear the area around the airfield the first Ju 52/3m transports could come in to land on the uneven runway and the nearby flat coastal strip. The landing area was under constant enemy artillery fire and within a few days the Maleme airfields and its surrounds looked like an aircraft cemetery.

In close cooperation, the paratroops and mountain troops gradually moved inland from west to east, reinforced by newly-landed formations, securing the island by 1 June 1941.

During this time the Royal Navy was present in force, first in support of the defending British troops and then assisting the evacuation. The British cruisers and destroyers operating around Crete were in turn the targets for the Luftwaffe bombers and Stukas of I and II/LG 1 and St.G 1 and 2 who managed to sink nine warships and ten freighters.

The total personnel losses in fighting for Crete amounted on the German side to 1350 dead and no less than 2700 missing, while the official British casualty figures list 1750 dead and 12,250 POWs. [Thanks to 'Ultra' codebreaking operations, the British command under General Freyberg, VC, was fully informed of German plans and strength – the first time such a situation arose in the war. Crete was defended by 42,640 troops (32,112 British and 10,258 Greek), of which 18,600 were evacuated – official figures. *Tr.*]

For the German paratroops these losses represented a bloodletting that painfully indicated the risks involved in an airborne landing among prepared enemy opposition.

In the meantime, quite unnoticed and under the tightest secrecy wraps, action was taken from mid-May 1941 to carry out the Führer's Directive No. 30. To support an anti-British uprising in Iraq, a special Luftwaffe unit known as *Sonderkommando Junck* was formed. Directly subordinated to the Luftwaffe High Command, it consisted of 4./ZG 76 with Bf 110D heavy fighters, 4./KG 4 with He 111H bombers and the *Transportstaffel Rother* with Ju 52/3m and Ju 90 aircraft. With their crosses painted over with Iraqi markings, this small formation was based at Mossul and carried out bombing raids and low-level attacks on British airfields at Habbaniya and Shaibah and troop encampments around the Iraqi capital. However, the British quickly managed to get the better of this effort so that the German aircrews had to be pulled back to Greece by the first week in June 1941. During these transport and transfer flights the Luftwaffe made use of French military airfields in Syria as intermediate landing grounds.

Still affected by the defeat in Greece the British divisions in North Africa could not withstand the advance of the newly formed *Deutsches Afrika Korps* (DAK) under General Erwin Rommel. As a result, nearly the whole territory given up by the Italians right up to the Egyptian border was regained during the German counter-offensive which began on 31 March 1941, supported by German and Italian squadrons.

Thanks to the occupation of Crete its airfields at Iraklion and Kastelli could now be used by Luftwaffe bomber *Gruppen* and (F)-*Staffeln* previously based in Greece. The formations thus able to support the fighting in Cyrenaica were I/JG 27, III/ZG 26, I, II and III/LG 1, II/KG 26, III/St.G 1, II/St.G 2 and 2.(F)/123, comprising the nucleus of X *Fliegerkorps* in the Mediterranean area. In the meantime, Operation 'Barbarossa' was making full demands on the bulk of Luftflotten 1, 2, 4 and 5.

Top: *Based at Plovdiv in Bulgaria with Ju 88As, I/LG 1 was one of the Luftwaffe bomber formations that prepared the way for the army from the very beginning of the Balkan campaign. This photograph, taken on 5 April 1941, shows a Ju 88A–5 returning from a bombing raid on the Metaxas Line.*

Centre: *On 16 April 1941, while taxiing to his take-off position, the pilot of this fully bombed-up Ju 87B–2 'fabricated' a classic undercarriage break. The aircraft operating in the Balkans were identified by yellow engine cowlings and rudder surfaces. This particular Ju 87B–2 belonged to Stab II/St.G 77 based at Agram.*

Bottom: *This Bristol Blenheim I of a Yugoslav bomber squadron at Nis fell victim to a low-level attack by Bf 110s of II/ZG 26. The 100 or so Hurricane I and Bf 109E fighters of the Yugoslav Air Force were destroyed at their bases or in aerial combat during the first days of this campaign.*

Top: Hptm *Wolfgang Lippert, commander of II/JG 27, has just landed with his Bf 109E–4 at Larissa in Greece. Camouflage nets stretched between the trees hide the parked aircraft from RAF reconnaissance.*

Centre: *This Bf 109E–3 of 5./JG 77 broke its undercarriage during a skidding landing at Almiros on 22 April 1941. The entire I/JG 77 had transferred from Brest-South to south-eastern Europe on 29 March 1941.*

Bottom: *During the time of the Balkan campaign the bulk of the Greek fighter squadrons was equipped with obsolete PZL–24 high-wing monoplanes bought from Poland before the war. The Royal Greek Air Force held out against the Luftwaffe for only a few days.*

Top: *In May 1941 the former civil Junkers G 38 'Hindenburg' airliner was used as a troop transport to Athens-Tatoi. This large transport aircraft flew in mainly mountain troops to Greece before Operation 'Merkur'.*

Centre: *A tractor drags two 1000kg (2205lb) bombs on a wooden sledge to the aircraft parking area at Eleusis. Preparations for the assault on Crete began immediately after the occupation of the Greek mainland.*

Bottom: *JG 77 armourers are loading up the 60-round drum magazines of the 20mm MG–FF cannon fitted in the Bf 109E wings. During the invasion of Crete JG 77 flew both escort and interception sorties from its forward landing ground at Molaoi in the Peleponnesus.*

Top left: *Towing a fully loaded DFS 230 glider, the Ju 52/3m is only minutes away from the coast and the open sea – destination Crete.*

Top right: *This DFS 230 assault glider was set on fire by enemy infantry weapons after landing on the strongly defended island. Many DFS 230A gliders crashed while landing on the stony, uneven terrain of Crete.*

Bottom: *On the evening of 19 May 1941, German air landing troops prepare their weapons and assault gliders for Operation Merkur on the following day.*

Top: *The airborne landing on Crete begins. While their comrades are still floating down to earth under their parachutes the first wave of paratroops are assembling west of the Maleme airfield. This action photograph also shows two-wheeled drop containers with extension axles.*

Bottom: *Attempts by Ju 52/3m pilots to avoid the Maleme airfield which was covered with crashed aircraft (left above in the picture) and set down on the narrow coastal strip ended with crash landings in the shallow water.*

Top: *A Ju 52/3m carrying a full load of mountain troops is on its way to the fiercely disputed island of Crete. The aircraft displays the characteristic yellow Balkan area colouring on its engine cowlings and rudder.*

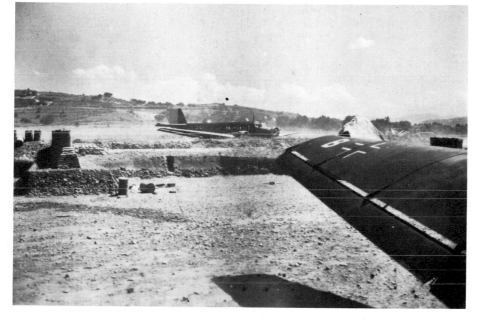

Centre: *After touching down, an 'Auntie Ju' feels its way carefully along the Malemes perimeter. One of its predecessors had attempted an outside landing and sheered off its undercarriage, although the crew and its passengers got away with just a few bumps.*

Bottom: *The approaching aircraft overfly some DFS 230 gliders that had landed earlier leaving skid marks from their plough-type brakes. The DFS 230A on the left has lost most of its port wing.*

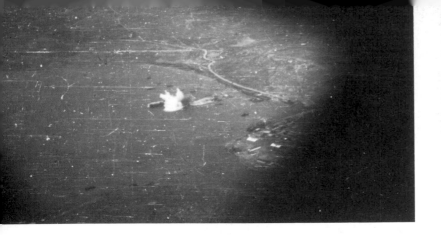

Top left: *A bomb hit on a British freighter in Suda Bay on 27 May 1941. The LG 1 together with St.G 1, 2 and 3 were out every day during the struggle for Crete, attacking supply vessels and warships of the Royal Navy in the waters around the island.*

Top right: *Gen Lt* Bruno Bräuer, *commander of Fj Rgt 1 (1st Parachute Rgt) and later fortress* Kommandant *of the island, jumped over Crete together with his men. Postwar, he was accused of war crimes, sentenced to death and executed in Greece in 1947. During the assault on Crete many wounded German paratroopers were massacred by local civilians and groups of Greek soldiers, which led to later repression – the usual story in such underhand warfare. Afterwards, both sides were 'right' – but the loser had to pay the penalty.*

Bottom: *A wrecked and burnt-out Ju 52/3m on the shore west of Chania. During the first days of this undertaking ships of the Royal Navy kept the Malemes airfield and the surrounding area under fire from their long-range guns.*

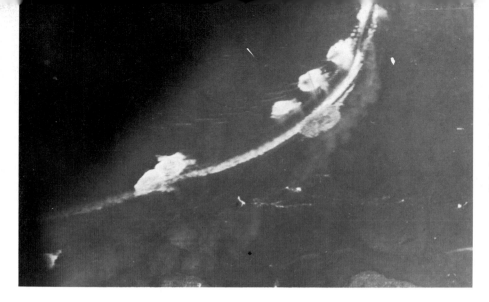

Top: *In the forenoon of 22 May 1941 Ju 88As of I/LG 1 commanded by Cuno Hoffmann succeeded in scoring several hits on the British cruiser* Niad *south of Helos, badly damaging her. The Royal Navy was active right until the evacuation of Crete.*

Centre: *A Ju 90B transport of* Transportstaffel Rother *shortly before take-off on a supply flight. In May 1941 this special unit had the task of flying in supplies to* Sonderkommando Junck (Special Detachment Junck) *operating in support of the anti-British revolt in Iraq.*

Bottom: *Overpainted with Iraqi national markings, the Bf 110D–3s of 4./ZG 76 were flown by their crews to Mossul. Early in June, after the collapse of the revolt, the Special Detachment Junck returned to Greece.*

127

Top right: *The Italian airfield at Catania in Sicily became one of the central bases for the formations of X* Fliegerkorps *which joined the air war in the Mediterranean area beginning January 1941.*

Centre: *One of the first Luftwaffe units based in North Africa was the 2.(H)/14 short-range reconnaissance squadron whose Bf 110Es provided target information for General Rommel's counter-offensive which had begun in April 1941.*

Bottom: *Stuka bombs are exploding in the surrounded and bitterly fought-over town and port of Tobruk. This fortress by the sea was under siege for six months, from July to December 1941, but could not be taken by Axis troops.*

Top: *A new unit arrives in the combat zone – JG 27, soon to be known as the 'Afrika-Geschwader'. The Bf 109E–4/N of* Oblt *Franzisket, adjutant of* I/JG 27, *during the transit flight from Catania to Castel Benito in April 1941. Note the 'Africa' insignia of JG 27 on the engine cowling and the 300 ltr (66 Imp. gall.) drop tank under the fuselage.*

Centre: *Ain el Gazala, a desert airfield, became the home of 3./JG 27 in May 1941. The* Staffelkapitän Oblt *Homuth is seen here with the squadron mascot, a young 'flying dog', while* Fw *Josef Kraus, one of the cadre pilots of the unit, leans against the cockpit.*

Bottom: *This Vickers Wellington fell victim to the Flak of* Afrika-Korps *on 25 March 1941. During that period the RAF bombers attempted to disturb the German–Italian assembly areas and supply routes.*

8. War on the Eastern Front

The military situation in the Mediterranean, France and the Scandinavian area in June 1941 showed that the prospects for realizing Hitler's political goal, the destruction of Bolshevism in Eastern Europe, were hopeful; the Germans reckoned that, with a timely start to the offensive and using tactics similar to those successfully employed in previous campaigns, it would be possible to reach the Volga beyond Moscow, and the Caucasus in the south, before the onset of winter.

The plans for Operation 'Barbarossa', in preparation since the end of 1940, were now to be put into effect as quickly as possible, starting on 22 June 1941. With the allied Finland in the north and the Hungarian and Romanian partners in the south the operational situation was very favourable for an offensive on a front stretching from the 70th to the 48th degree of latitude.

The commanders of Luftflotten 5 (*Gen* Stumpff) in the north, LF 1 (*Gen Oberst* Keller) in the East Prussian area and northern *Generalgovernment* (occupied Poland), LF 2 (GFM Kesselring) in southern *Generalgovernment* and LF 4 (*Gen Oberst* Löhr) in Slovakia, Hungary and Romania, had received orders outlining the forming-up deployment on 13 January 1941. According to these orders, LF 1, 2 and 4 had 11 (F)-*Staffeln*, 90 bomber *Staffeln*, 59 fighter *Staffeln*, 18 replacement fighter *Staffeln*, 6 *Zerstörer-Staffeln* (heavy fighters), 3 replacement heavy fighter *Staffeln*, 24 Stuka-*Staffeln* and 3 coastal-aviation *Staffeln* at their disposal. In addition, 79 short-range and Panzer-reconnaissance *Staffeln*, 14 long-range reconnaissance *Staffeln* and 11 courier *Staffeln* were directly subordinated to the three Army Groups.

Initially, LF 5 was intended as an operational reserve, leaving LF 3 alone to continue aerial warfare against Britain and shipping in the Atlantic, as well as providing defence against British penetration of the air space over German-occupied territories and Germany itself.

At 0300 hrs on 22 June 1941, at first light, the first German bomber *Geschwader* roared across the German-Soviet demarcation line, their targets – the airfields. It was a matter of hitting Soviet air power as hard as possible on its advanced bases at the very beginning and, in fact, some 800–850 Soviet aircraft were destroyed on the ground on that very first day. Despite that, swarms of I–16 and I–153 fighters threw themselves against the German *Geschwader* during their second raid. In the afternoon began the first attacks by Soviet SB–2 and DB–3 bombers on advanced Luftwaffe airfields. It was an uncanny experience: the Soviet bombers appeared as if on parade, in strict formation flight, holding their course through the anti-aircraft fire and Luftwaffe fighter attacks. The results were terrible: one after another the portly twin-engined Soviet bombers fell victim to German defences. By evening the total number of Soviet aircraft destroyed on the ground and in the air had risen to at least 1800 – a one-day balance unique in the entire war.

The fast advance of German army formations during the first weeks also seemed to indicate that Hitler's reckoning could well be right – to reach the planned operational goals of this campaign by November. Superior in the air, the Luftwaffe could give the ground forces all the support they needed – with the result that in the end this support became so indispensable that the OKL (High Command of the Luftwaffe) gradually began to lose sight of its real task, strategic air warfare.

Combat reports of the bomber and Stuka *Gruppen* mention railway stations, railway lines, bridges, railway junctions, troop assembly areas, artillery positions, bridgeheads and many other targets – which are always connected with immediate events on the ground and seldom further than 100 km behind the front lines. Power stations, industrial establishments, supply installations and supply centres are rarely mentioned in target data lists for 1941/42. Besides an appropriate perception of its role, the Luftwaffe also lacked aircraft types of sufficient range to carry out long-term tasks involving a massive concentration of effort. In fact, Operation 'Barbarossa' was planned by the German command on the same lines as the previous 'Blitzkrieg' campaigns, without regard to the depth and breadth of Russian air space.

The first signs of a crisis began to show in October 1941, with the start of the mud period, and early in December the crisis was there in earnest. The Luftwaffe raids also changed: instead of the goal of the offensive, Moscow – which had been the recipient of some 'disturbance raids' in October and November, other air raids in November were aimed at the Gorky industrial centre, as well as the towns of Rzhev, Smolensk, Orsha, Bobruisk, Zhitomir, Kiev, but especially Leningrad, which had to suffer a hail of German bombs. With the exception of KG 26, KG 30 and KGr.100 all other Luftwaffe bomber *Geschwader* had in the meantime been transferred to the Eastern Front, and EGr.210 had become SKG 210, a high-speed bomber *Geschwader*. Equipped with Bf 110s, it flew surprise attacks on Soviet troop and tank assembly areas and approach roads behind the front lines together with II(Schl.)/LG 2, the only assault/ground-support *Gruppe* of the Luftwaffe.

The *Stukageschwader* 1, 2 and 77, still equipped with the Ju 87B or the longer-ranging Ju 87R, remained the 'water carriers' of the Army divisions. They also had to carry out the most moves of all Luftwaffe formations. In some cases the areas of concentrated main effort were hundreds of kilometres apart.

By 9 September 1941 Leningrad was completely surrounded, leaving only the Ladoga lake as a possible supply route. Just two months later, Operation '*Taifun*' (Typhoon), the German attack on Moscow, got bogged down in Russian winter a few kilometres from its goal. Apart from that, there were also signs of Soviet aerial superiority in this section. The front lines of the German Army Groups North and Centre had become fixed when strong Soviet counter-attacks began early in December 1941, gradually forcing back the German lines. The Luftwaffe could help only intermittently because it lacked pre-heating installations for the frozen aircraft engines; the winter was very harsh. To relieve Leningrad, the Soviets achieved several breakthrough penetrations south of the Ilmen lake in January 1942, which resulted in the 'cauldrons' of Kholm with 5500 men of the *Kampfgruppe* (combat group) Scherer, and some 95,000 men of II Army Corps surrounded in the Demyansk area. A number of Luftwaffe

transport *Gruppen* were quickly deployed to supply these surrounded troops and *Oberst* Fritz Morzik was appointed *Lufttransportführer* (air transport leader) to coordinate these efforts. The air-supply sorties began on 20 February 1942 with seven transport *Gruppen*. In the 'Cauldron' of Demyansk there were two provisional airfields, at Demyansk itself and Pyesti, where the transport aircraft could unload their cargoes.

The daily supply quota was set at 300 tons, an impossible target for the 220 available transport aircraft, of which only an average of 30–50 per cent were serviceable at any one time. To help out, another four transport *Gruppen* were made available from LF 4 area, followed early in March 1942 by another four transport *Gruppen* formed under the aegis of the Chief of Luftwaffe Training. The starting bases for this air transport fleet were the airfields at Pskov-West, Pskov-South, Riga, Daugavpils (both in Latvia), Korovye-Selo and Tulebiya. Due to the destruction of its small landing ground and the diminished size of the 'cauldron', Kholm had to be supplied by air drops and one-way trip cargo gliders.

Both air-supply operations were successfully carried out until June 1942 when the surrounded German troops were relieved. Unfortunately this led to the erroneous assumption that it would also be possible to supply larger formations from the air. At the same time it was overlooked that the transport units had lost 262 aircraft and 385 aircrew – including most of the instructor personnel from flying training schools impressed into the last four transport *Gruppen*.

Right through 1942 the Chiefs of the three Luftflotten had their hands full supporting according to the proven pattern the armies and Panzer Groups from the air in their position warfare, withdrawal moves and break-through battles. Points of massed concentration for the flying formations developed on the southern sector of the Eastern Front during the siege of the Sevastopol fortifications (January–July 1942), the advance of the Army Group A into the Caucasus (July–November 1942), the advance towards the Central Don area (June/July 1942) and, under the code-name 'Braunschweig', during the assault on Stalingrad.

Despite impressive victory scores achieved by individual fighter pilots, aerial superiority had already been lost over many sectors of the frontline. New Soviet aircraft types including the MiG–3 and LaGG–3, and the first Lend-Lease supplies from the Western Allies, all helped the Soviets to stand their ground against the Luftwaffe, and more.

In recognition of the situation, *Schlachtgeschwader* 1 (close support/ground attack unit) had been formed in the strength of two *Gruppen* from II(Schl)/LG 2 in January 1942 and used operationally as a mixed formation (being equipped with the Bf 109E fighter-bombers and Hs 123 and Hs 129B close support aircraft) on the southern sector of the Eastern Front.

The operational activities of LF 5 in the Norwegian and Finnish areas had only a limited influence on the course of the air war on the Eastern Front. Until August 1941 the operational sorties of the He 111Hs of I/KG 26 and Ju 88As of KG 30 had been exclusively nocturnal bombing raids on British ports and daytime attacks on shipping along the eastern shores of the British Isles; however, during the following months the bombers were gradually moved further north for other tasks. Their new bases were at Alta, Bodö, Banak, Kirkenes and Bardufos in northern Norway, and Petsamo in Finland, and their targets were the Allied ships plying with war material for the Soviets around the North Cape, the supply port of Murmansk and the Murmansk railway line. The reconnaissance and weather information for these operations in the far north were obtained by the crews of *Wekusta* 5 (weather reconnaissance *Staffel*), 1.(F)/120 and 1.(F)/124. In the meantime, IV(Stuka)/LG 1, in close cooperation with the Finnish Army Corps, supported the battles in Karelia (Operation '*Silberfuchs*' = Silver Fox) all the way up to the White Sea. The fighter force was provided by I and IV/JG 77 and the 'heavy fighter' *Staffel* of JG 77 – units, which from January 1942 formed the nucleus of JG 5, the '*Eismeergeschwader*' (Polar Sea *Geschwader*).

Until March 1942 only lone-sailing freighters and tankers appeared in the success reports of these Polar-based Luftwaffe bomber crews, although Allied convoys had been sailing along the drift-ice line, transporting war materials to Murmansk and Arkhangelsk since summer 1941. After such a convoy had been spotted for the first time on 5 March 1942 by a German reconnaissance aircraft, *Gen Oberst* Stumpff began sending combat aircraft at his disposal – including the Fw 200Cs of I/KG 40 – on deliberate attacks on these supply routes. From 2 to 10 July 1942 the objective was Convoy PQ 17, which lost 24 ships while sailing around the North Cape. Two months later, beginning 13 September, Convoy PQ 18 lost 13 ships to Luftwaffe bombers and torpedo-bombers in just two days. But such successes were not common.

Due to weather conditions up north during the winter 1942/43 there were few longer operational periods for these aerial activities, and this was the time when a disaster for the German Sixth Army and the Luftwaffe transport aircraft was in the making down on the southern sector of the Eastern Front: the tragedy of Stalingrad.

On the basis of experience gained while supplying the Demyansk 'cauldron', the Commander-in-Chief of the Luftwaffe, *Reichsmarschall* Hermann Göring had guaranteed the air supply of the beleaguered Stalingrad Army (which had been surrounded since 22 November 1942) in Hitler's presence. According to orders, a day later *Gen Ltn* Fiebig in his new capacity as *Luftversorgungsführer Stalingrad* (air supply leader) began to assemble all available transport aircraft at Tazinskaya to fly in the daily requirement of 300 tons. After a rather slow build-up it was only on 19 December 1942 that these supply flights got near the required amount – 290 tons. With the loss of the foreward airfields at Tazinskaya (on 24 December 1942) and Morozovskaya (early in January 1943) the flight path to the remaining landing grounds at Pitomnik and Gumrak was increased to 350 km – which in turn reduced the amount of supplies carried. Provided with special powers, *Generalfeldmarschall* (GFM) Milch had taken over the post of air supply leader Stalingrad on 16 January 1943 but, despite great efforts even '*Sonderstab GFM Milch*' did not succeed in flying in the necessary amount of supplies into the 'cauldron'. The airfield at Pitomnik, covered with the wrecks of crashed aircraft, had to be given up on the day GFM Milch had taken over; that at Gumrak was abandoned on 22 January 1943. The new temporary landing field at Stalingradsky was in use for just one day before Soviet artillery fire stopped all air traffic there as well. Between 24 January and 3 February 1943 food and ammunition for the German defenders of Stalingrad could only be supplied by parachute drops – and of course this also meant the end of casualty evacuation. Added to the negative balance of the loss of the entire Sixth Army were the 1000 or so aircrew and 488 transport aircraft – far too high a price to pay for this failed experiment. Stalingrad was to remain as 'writing on the wall' for the duration of the war.

The year 1943 began with bitter defensive battles on all sectors of the Eastern Front and the withdrawal of German lines, which at times laid claims on the considerably weakened transport *Gruppen* again. The terms 'straightening of the front line', 'clearing-out of break-ins' and others similar, which began to appear in the German High Command communiqués from spring 1943 onwards, disguised a lost German initiative, which resulted partly from the demands of all-out war on several fronts, and partly because of the material superiority of the opposition. Under these conditions the Luftwaffe command had adopted a number of successful recipes used by the 'other side' and introduced them in the various Luftflotten. Thus, for instance, based on the pattern of the Soviet 'sewing machine' aircraft tactics the first German 'night disturbance' *Staffeln* appeared early in 1943. Later on, they were organized into special NSGr (*Nachtschlachtgruppen*, night assault groups), equipped with various training aircraft, such as the Ar 66, Go 145, Fw 58 and He 46, sprayed black. These aircraft had the task of causing unrest behind the Soviet lines by attacking every light glimmer with light bombs or other kind of 'special ammunition' – hand grenades, demolition charges and suchlike.

In imitation of the heavily armoured Soviet 'Zementer' ('Concreter'), the I1–2 assault aircraft, the Luftwaffe had introduced the twin-engined Hs 129B ground support/attack aircraft, armed with large-calibre cannon, in May 1942. The first aircraft of this type were delivered to II/SG 1. A fighter unit, JG 51, also had a 'tank destroyer' *Staffel* equipped with the cannon-armed Hs 129B.

The successes achieved with the new assault and 'tank hunting' tactics were such that they practically demanded the extension of these special formations. From April 1943 the first Hs 129B-*Staffeln* of the new SG 2 began flying low-level attacks on Soviet tank concentrations and road traffic. A special 'Experimental Command for Combatting Tanks' tested the suitability of the Ju 87 and Ju 88 armed with large-calibre cannon (37 mm and 75 mm) for this task, becoming the nucleus of the 10.(Pz)-*Staffeln* attached to each Stuka-*Geschwader*. The successful attack tactics with the Ju 87G evolved by *Oberst* Rudel made this officer, who was awarded the highest German decorations for bravery, an internationally known 'tank destruction' expert.

The Luftwaffe bomber arm too had become interested in the possibilities of effective low-level attacks on marching columns of enemy troops and railway traffic with heavily armed 'destroyer' aircraft. The concentrated fixed fire power of the Ju 88C made this 'all-rounder' into an ideal 'hedge-hopper' with which to equip the '*Zerstörer*' ('destroyer' or heavy fighter) and 'Eis.' (= *Eisenbahnbekämpfungs*-, railway attack)-*Staffeln* formed in bomber *Geschwader* from summer 1942 onwards. With the introduction of these kinds of army support, which were not new but appeared relatively late with the Luftwaffe, the largest part of what hitherto had been the 'flying artillery' began operating elsewhere.

New on the Eastern Front were also the *Nachtjagdschwärme* (night-fighting flights) stationed within the areas of the individual Luftflotten. Employed from about autumn 1942 onwards against the guerilla-landing and supply flights – which had become like a plague in the meantime – these specialized night-fighting flights were later misused in different roles during the great withdrawal movements and practically annihilated. It was only the formation of NJG 100 and NJG 200 (each of just one *Gruppe* strength) in August 1943 that made it possible to start combatting the nocturnal Soviet penetration flights with some success again.

There was only one more concentrated effort by the Luftwaffe in the East, on the scale of the preliminary raids on Britain before the planned invasion in autumn 1940. Operation '*Zitadelle*' (Citadel), planned for summer 1943, was intended as a large-scale 'pincer attack' to cut off and destroy the Soviet forces concentrated in the Kursk 'bulge'. Including formations transferred from other sectors, *Gen Maj* Seidemann (VIII *Fliegerkorps*), commanding the Luftwaffe effort, had a force of no less than 950 aircraft to support '*Zitadelle*'. The attack began in great style early on 5 July 1943. During the period to 19 July, when the whole operation had bogged down and had to be discontinued, the Luftwaffe managed to fly up to 3000 sorties per day, a performance that came near its earlier achievements.

The step-by-step, deliberate and controlled withdrawals of the frontline after the Stalingrad catastrophe and throughout 1943 were no longer possible the following year. The increasingly numerous guerilla groups, distributed all over the hinterland, endangered all planned operations and also interferred with behind-the-lines communications and supply routes. As a result, in addition to frontline duties, the Luftwaffe formations also had to participate in anti-guerilla warfare, flying reconnaissance sorties and attacking guerilla camps and supply bases behind their own frontlines.

Then there was the Crimea peninsula, which had been cut off from the mainland and had to be supplied by air transport units – now partly equipped with large-capacity aircraft – throughout the spring of 1944. From their bases at Odessa and Kirovgrad, the transport *Gruppen* would fly across the Black Sea to the landing grounds at Karankut and Saki on the Crimea. Most of these supply operations had to be flown without fighter escort, which was seldom available. In line with the withdrawal from Crimea beginning April 1944 these supply flights became evacuation sorties, eventually carrying 21,500 soldiers back to the mainland. Some of the heavy war material could be brought back across the Black Sea by Siebel ferries operated by the Luftwaffe.

In this time of daily withdrawals the Luftwaffe managed to achieve another remarkable success against an American Eighth Air Force bomber formation which, with its fighter escort, had continued eastwards after bombing Berlin and landed on the Soviet airfields at Poltava, Mirgorod and Piryatin. This 'shuttle' operation was spotted by a German contact aircraft and nearly 200 Luftwaffe bombers raided Poltava on the night of 21/22 June 1944. Assisted by 'pathfinders' and 'illuminators' of KG 4, the He 111H bombers of KG 53 and 55 succeeded in destroying 47 four-engined American bombers and 15 fighters in high- and low-level attacks. The already very weakened Luftwaffe had 'shown her teeth' once more.

With the frontlines approaching the German borders and the accelerated reduction of Luftwaffe bomber *Geschwader* in favour of newly-formed fighter and ground-support *Gruppen* the main emphasis late in 1944 moved to tactical support of the ground troops in daytime and nocturnal assault operations. Aircraft of the 'new generation', such as the Ar 234 and Me 262, did not become operational on the Eastern Front.

During the final battles against the onrush of Soviet forces during December 1944/January 1945 most of the Luftwaffe units remaining in the East were already operating from bases on German soil. Soon, the Luftwaffe units defending Germany stood back to back with the remaining Luftwaffe formations in the East: all-round defence had become an unalterable necessity.

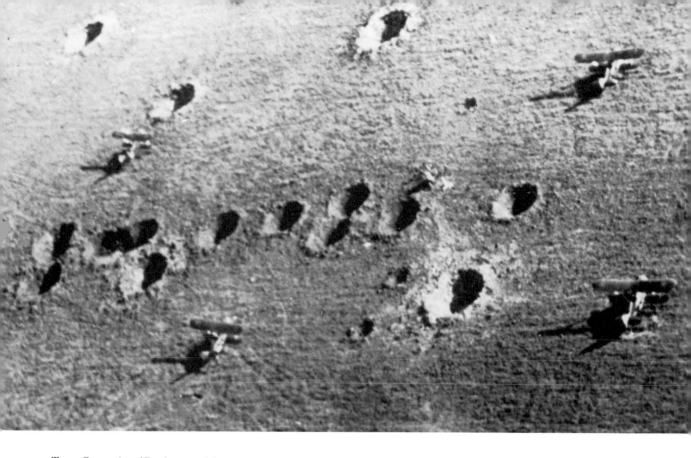

Top: *Operation 'Barbarossa' began with massive air raids on Soviet airfields in the morning of 22 June 1941. In this case it was clearly a matter of having used the wrong type of bombs: the aim was good, but the bombs exploded too deep in the soft ground creating imposing craters but hardly any splinter effect.*

Bottom left: *Shortly after its arrival from France, I/KG 53 'Legion Condor' was flying its first operation in the East, taking off from the advanced landing ground at Grojec near Warsaw. Here He 111H–5 A1+JK is pictured returning to base.*

Bottom right: *'Devil's eggs' – the small but deadly SD 2 anti-personnel splinter bombs being unloaded from their transport boxes. Fitted with two winglets, this 2kg (4.4lb) bomb would flutter to the ground and then explode into about 300 small splinters. Its preferred use was in low-level flight against marching columns of enemy troops.*

Top: The 8.(H)/32 Army reconnaissance squadron operated with Hs 126Bs in the northern sector of the Eastern Front, assisting the Eighth and Eleventh Armies in the advance towards the area Liepāja-Daugavpils-Šiauliai in Soviet-occupied Latvia and Lithuania. The angled coloured lines painted on the fuselage, known as the so-called 'Prenzlau stripes' (from the pre-war training base) were aiming lines to assist the observer.

Centre: The Soviet Air Force was hit really hard on the ground and in its confused retreat also left many intact aircraft on advanced landing grounds. Here are two of a larger group of I–16 Type 17 'Ratas' armed with two 20mm ShVAK cannon in the wings and two 7.62mm ShKAS synchronized machine guns in the fuselage. Lutsk airfield, July 1941.

Bottom: By the end of June 1941 Ofw Sepp Wurmheller of 5./JG 53 had 18 confirmed victories on his account. Later transferred to JG 2, 'Richthofen' Wurmheller became one of the most successful fighters in the West and by early summer 1944 had been promoted to the rank of Major. He was killed in action on the Invasion Front on 22 June 1944: during a furious dog-fight his Fw 190A collided with that of his wingman.

Top: *The Fi 156 Storch evolved into an ideal courier and liaison aircraft for use by the staffs of* Luftflotten *and* Fliegerkorps *in the East. A Fi 156C–2 is seen here following a country road in low-level flight behind the front lines.*

Centre: Gen Lt *Robert Ritter von Greim, a Pour-le-Mérite fighter pilot of the First World War, commanded the V* Fliegerkorps. *Special* Flugbereitschaften *(Flight Standby Detachments) equipped with He 111s and Do 17s were at the disposal of higher staff for long-distance flights – in this case a He 111 piloted by an* Oberfeldwebel.

Bottom: *The Chief of the Luftwaffe General Staff,* General der Flieger *Hans Jeschonnek, arrives for an informative visit to the headquarters of* Luftflotte 2 *(Kesselring) at Minsk in September 1941. After a series of violent quarrels with Göring, Jeschonnek was to commit suicide on 19 August 1943.*

Top: *Among the flying formations allied to Germany were the Slovak fighter squadrons. Equipped with obsolete Avia B 534 biplanes, their role was to secure lines of communications. In 1942, some of the Slovak fighter units were re-equipped with the Bf 109.*

Centre: *Hungarian fighter pilots joined the Russian conflict in June 1941, flying their Bf 109E and Fiat CR.42 fighters with keen elan. The Reggiane Re.2000 Falco I fighter in the photograph was delivered to operational units beginning summer 1943.*

Bottom: *A small Latvian flying formation equipped with the stout Anbo IV high-wing monoplanes took over courier and spare-part supply flights to Luftwaffe* Geschwader *operating in the northern sector of the Eastern Front.*

Top: *An aerial reconnaissance photograph of the Bryansk railway station taken the day after a bombing raid by I/KG 53 'Legion Condor' on the night of 29/30 July 1941. This target effect picture reveals many hits on railway tracks and burning freight trains and storage buildings.*

Bottom: *After the occupation of Orsha the damaged runways and buildings of the Orsha-South airfield were repaired and the airfield became an operational base for Luftwaffe bomber and Stuka formations.*

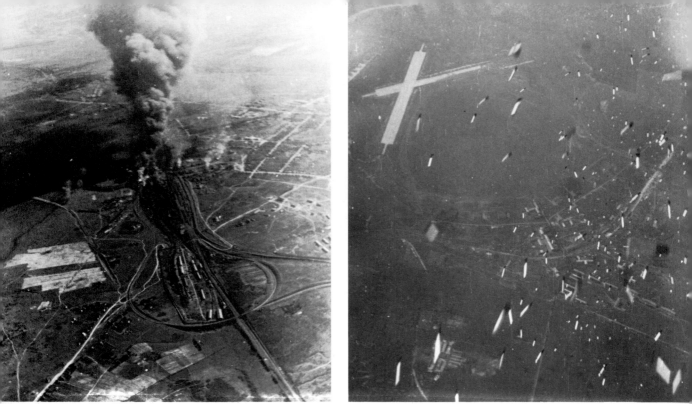

Top left: *The I/KG 53 continued its attacks on railway installations on 31 July 1941 when the entire* Gruppe *raided the railway junction of Vyazma. A growing column of smoke over the target confirms the effectiveness of the raid.*

Top right: *A rain of small incendiary bombs from a container falls on the hangars and barracks of the Soviet airfield at Konotop on 24 August 1941. This picture was taken from the ventral gun position of an He 111 of I/KG 53.*

Bottom: *While taking off on an operational flight from Jürgenfelde late in July 1941, a Ju 88A–5 of 5.(F)/122 long-range reconnaissance squadron suddenly ground-looped and caught fire. Fortunately the crew managed to get out in time from the fully tanked-up aircraft.*

Top: *An abandoned Soviet ANT–7 on an airfield near Orsha. [ANT–7 was the factory designation of the Tupolev Kr–6 heavy escort fighter – and the R–6 reconnaissance aircraft of the early 1930s – long since obsolete by 1941. A number of R–6s were still in service on second-line duties in summer–autumn 1941, mainly as transports. Tr.]*

Centre: *A trio of Bf 110Es of 1./SKG 210 during transfer flight to Rogoznitska. The crews of this Gruppe attacked mainly airfields and bridges with their guns in low-level flight.*

Bottom: *In August 1941 Major Arved Crüger (left) took over command of SKG 210 from Major Walter Storp in Seshchinskaya. Until September 1942 when he was ordered to take over command of KG 6, Storp served on the Operational Staff of the Luftwaffe.*

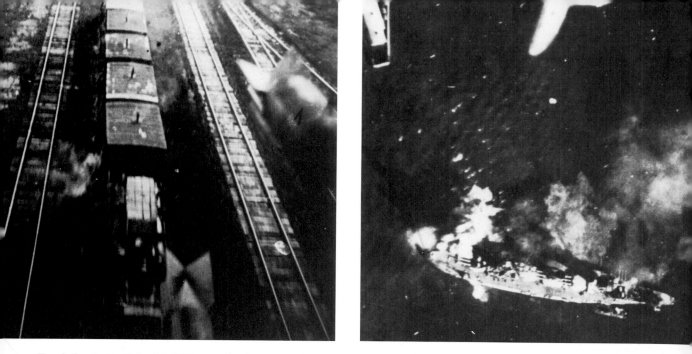

Top left: *A Ju 88A of 5./KG 54 unloads its 'eggs' in very low-level attack on a moving Soviet transport train on the route Kubyansk-Lissichansk on 30 September 1941. Only delayed-action bombs could be used from such low altitudes.*

Top right: *The target for III/St.G 2 'Immelmann' on 23 September 1941 was the Soviet warships at their base in Kronshtadt near Leningrad. Despite very strong anti-aircraft fire Stuka bombs scored hits and disabled the battleship* Oktyabrskaya Revolutsiya.

Bottom: *Another sample of precision bombing by Stukas is this hit on a pioneer-built bridge at Bronnitsi, achieved by aircraft of St.G 77 in August 1941. The permanent bridge in the background had been destroyed by aircraft of the same unit a few days previously.*

Top left: Oblt *Lothar Lau, the Technical Officer of* St.G 2 'Immelmann', *reports the combat strength of his unit to the* Fliegerkorps. *Ryelbitsy, August 1941.*

Bottom left: *On a higher level,* Generalfeldmarschall *Erhard Milch in his function as* Generalluftzeugmeister *(Chief of Supply and Procurement) informs himself about the technical situation of the Luftwaffe units in the East during a front-line journey in spring 1942.*

Top right: Oberst *Werner Mölders, the newly appointed Inspector of Fighters, pays a visit a fighter* Gruppe *based at Shatalovka in August 1941.*

Bottom right: *The highly decorated fighter ace Major* Hermann Graf *was also the goalkeeper of the leading Luftwaffe football team, the 'Roten Jäger' (Red Hunters). He is seen here talking to the trainer of this elite eleven (and post-war trainer of the Federal German football team) Sepp Herberger in May 1942.*

Top left: *A direct hit by a Soviet anti-aircraft shell had ripped away almost half of the vertical tailplane of this Ju 87B–2 of 6./St.G 1. The 6G + GT probably would not have made it back to base if the Soviets had followed up with a fighter attack.*

Top right: *Three officers of II/JG 54 'Grünherz' at Kalinin in October 1941. From left:* Oblt *Hans Philip,* Oblt *Wolfgang Späte and the* Gruppe *commander Dietrich Hrabak. Hans Philip had achieved 206 victories (177 East + 29 West) when he was killed in action in 1943; Wolfgang Späte, after a successful career as a fighter pilot (99 victories), was appointed in December 1944 to command JG 400, the only unit to fly the rocket-powered Me 163 Komet in combat; while Dieter Hrabak (total of 125 victories) ended the war as last commander of JG 54 in Courland/Latvia.*

Bottom: *Next to the I–16 'Rata' the I–153 'Chaika' (Gull) was the foremost standard equipment of the Soviet fighter force in 1941. Not very fast, the I–153 was extremely manoeuvrable and a well-versed pilot could easily get out of his follower's sight by a steep opposite turn. [The Shcherbakov I–153 was also notable as the last biplane ordered in series production anywhere, and the only operational aircraft of that type with a retractable undercarriage. Tr.]*

Top: *Supplying the 'cauldron' of Kholm in February 1942: in this area, surrounded by Soviet troops, there were no operational landing fields and supplies had to be dropped by parachute or flown into the 'cauldron' by cargo gliders. Gotha Go 242 gliders being loaded up at Pskov-South, February 1942.*

Centre: *Rolling petrol tanks. Steel cylinders, 1.6m in diameter, strengthened by wide strips, could be rolled directly to the aircraft to speed up refuelling in Russian winter conditions. Some He 111 bombers were also impressed to fly in supplies to the Demyansk 'cauldron'. Pskov, February 1942.*

Bottom: *The thaw in March 1942 made flying operations much more difficult. The crews of the transport Gruppen had to repeat their trips across Soviet-held territory to Demyansk several times daily to ensure sufficient supplies for the surrounded II.Army Corps. The relative success of this air supply operation, the first of its kind and size in the history, culminating in the break-out of the surrounded troops, was to have dire consequences at Stalingrad in the following winter.*

Top: *The Dno airfield, south-west of the Ilmen lake, was used by 3.(F)/22 as base for reconnaissance flights over the northern sector of the Eastern Front with its Ju 88s and Bf 110s. The Bf 110C in winter camouflage was the machine flown by the Staffelkapitän Oblt Hermann Fischer.*

Centre: *The daily bombing raids on Soviet troop assembly areas, tank concentrations and supply routes by He 111Hs of 1./KG 53 'Legion Condor' contributed their share to the stabilization of the front lines in the northern sector.*

Bottom: *A shock for the intercepting fighter! To strengthen its defensive fire power the He 111H–6 was fitted with a fixed 7.92mm MG 17 built into the tail cone of its fuselage. This 'fright weapon' was fired by the pilot, who could keep an eye on his pursuer in a backward-facing sight.*

Top: *The snow-covered airfield at Luga with Ju 88Ds of 3.(F)/Aufkl. Gr.Ob.d.L. This special long-range reconnaissance Staffel was directly subordinated to the Commander-in-Chief of the Luftwaffe and, after rest and re-equipment at Orly near Paris transferred to the Eastern Front in February 1942.*

Centre: *During the crisis months in early 1942 casualties were occasionally flown back to Germany aboard temporarily impressed Lufthansa aircraft. Ju 90Z–2 D–AEDS taking on a load of seriously wounded soldiers at Siverskaya in February 1942.*

Bottom: *The Fi 156D flying ambulance version could carry one stretcher and one medical attendant in its narrow fuselage. The D–ELXV of* Sanitätsflugbereitschaft *17 (Flying Ambulance Standby Duty Detachment 17) undergoing engine change at Nikolayev late in February 1942.*

Top: *Carrying two SD 250 (551lb) splinter bombs on its fuselage bomb racks, a Bf 110F–1 of II/ZG 1 takes off from Kharkov in June 1942. The Bf 110F–1 was the bomber version of the Bf 110 Zerstörer.*

Centre: *From May 1942, Luftwaffe ground-support formations were reinforced with the heavily-armoured Henschel Hs 129B aircraft. Here a force-landed Hs 129B–1 of II/Sch.G 1 is put back on its feet by means of a tripod lifting jack.*

Bottom: *One of the new Soviet fighter types making its debut during the Great Patriotic War (as the Soviets called it) late in 1941 was the MiG–3. About 2000 machines of this type were produced for the Soviet Air Force between 1941 and 1943.*

148

Top: *The Go 242 large-capacity transport gliders began to find an increasing role on the Eastern Front in 1942. Here three Go 242A gliders in tow are seen overflying the Dnepr bend at Vitebsk. Thanks to its effective camouflage the left lower aircraft is hardly visible.*

Centre: *An edifice well known to every Luftwaffe airman in the East! One of the typical board-covered flight control buildings seen on most airfields in Russia. In front of it, a similarly typical 'Panje' carriage from the airfield 'duty centre' for especially urgent cases.*

Bottom: *One of the leading Luftwaffe fighter aces in summer 1942 was Major* Gordon Gollob, *commander of JG 77, with 107 confirmed victories to his credit. This photograph was taken at the Oktober airfield on the Crimea in mid-1942.*

Top: *The heaviest bomb used by the Luftwaffe was the 2500kg (5511lb) SB 2500 'Max'. Its explosive charge amounted to 1700kg (3748lb). Here is a 'Max' on its special carriage destined for KG 1 'Hindenburg' at Dno in spring 1942.*

Bottom left: *A most welcome replacement for the sedate Hs 126B short-range reconnaissance aircraft was the FW 189A, increasing numbers of which began to join the Army reconnaissance formations from March 1942 onwards. This is the T1+EH of 1.(H)/10 'Tannenberg' in April 1942.*

Bottom right: *Lt Ernst Ortegel with his crew survived several bad crashes while serving with the 4.(H)/10 Army reconnaissance squadron. In May 1943 his unit shared the Kharkov-North airfield with several other formations.*

Top: Generaloberst *Hans-Jurgen Stumpff, commander of Luftflotte 5, was awarded the Knight's Cross on 18 September 1941. The Luftflotte 5 combined formations based in Norway and Finland, controlled from headquaters in Oslo.*

Bottom left: *On 4 July 1942 He 115 floatplanes of 1./406 coastal squadron based at Tromsö/Norway participated in attacks on the convoy PQ 17 near the Bear Island. Here the squadron artist completes the success score marked on the fin of one of the aircraft.*

Bottom right: *Among the units based in the Arctic regions were also some coastal squadrons equipped with He 115 torpedo floatplanes. A red-and-white training torpedo is being lifted into the fuselage of an He 115B–1.*

Top: *Until the end of May 1941, KG 3 (known as the 'Blitzgeschwader') was subordinated to Luftflotte 2 and took part in raids on the British Isles. Transferred East, it was operational there from the very first day of 'Barbarossa'. After taking off from Shatalovka, a Ju 88A–4 of II/KG 3 is seen here on its way to attack targets in the Orel area on 23 October 1941.*

Centre: *Some Luftwaffe bomber formations on the Eastern Front included Ju 88C heavy fighters on detachment to combat Soviet locomotives, anti-aircraft gun positions and road traffic. Posing in front of their well-armed aircraft is a successful crew of 6./KG 3 engaged on such activities. From left: Fw M. Ringer (radio operator/gunner), Ltn O. Hansberg (pilot) and Fw G. Kimscher (observer). Orsha-South, May 1942.*

Bottom: *The yellow propeller hubs of this Ju 88A taxiing to its take-off position at Bagerovka on the Crimea in May 1942 identifies it from a distance as belonging to III/KG 51. Known from its insignia as 'Edelweiss-Geschwader', KG 51 was operational the year round mainly from bases in the Black Sea area and was equipped with Ju 88A–4 bombers.*

Top: *A closer look at a shot-down Soviet Ilyushin Il–2 ground-attack aircraft. Known to German troops as the 'Zementer' (Concreter), it was built around an armoured shell. At times swarms of these machines made life difficult for German airfields, road transport and artillery positions.*

Centre: *At the beginning of the Eastern Campaign special courier squadrons were formed and subordinated to individual Army Groups and Armies. The commanding general of Army Group A,* Generaloberst List, *thanks the commander of* Kurierstaffel A, *Oblt* Hermann Becker, *on a forward landing field in 1943.*

Bottom: *In summer, the films brought back by reconnaissance crews were dried out on large drums in the open. Here the photographic personnel of 3.(F)/22 at Dno in July 1942 puts on to reels film taken by an automatic camera.*

Top: *On 19 August 1942, the Sixth German Army was ordered to attack Stalingrad. Among the Luftwaffe aircraft supporting this offensive were the Ju 87B–2 Stukas of 6./St.G.1 which bombed targets in the area of advance on 23 August. Within St.G.1 the aircraft flown by* Staffelkapitäne *were distinguished by their pale yellow painted rudders.*

Centre: *The lessons learned from experience in the first hard winter on Russian soil were many. Hurriedly nailed together 'pre-heating huts' helped engine starting after a cold night in winter 1942/43.*

Bottom: *To master the masses of snow good use was made of captured Soviet caterpillar tractors pulling heavy triple rollers to flatten and harden the aprons (the areas in front of the hangars) and taxiing routes to the take-off runways. The available airfield personnel could have never managed to shift the daily fall of new snow.*

154

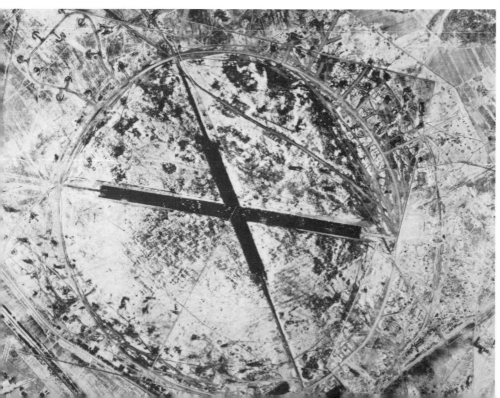

Top: *The declared target of the offensive, Moscow, was never reached by the German forces. An aerial photograph taken on 14 August 1942 shows the central section of the Soviet capital. The triangle formed by roads left of the Moskva river bend is the Kremlin area.*

Bottom: *The characteristic runway arrangement of the near-circular Seshchinskaya airfield, with only the two take-off/landing runways cleared of snow. This airfield was one of the most intensively used by the Luftwaffe in the East.*

155

Top: *The air supply of 'Fortress Stalingrad' begins. While the weather was still good in December 1942 it was possible to fly the Ju 52/3m transports in formation right into the surrounded area near the Volga river.*

Centre: *The two main landing fields in the Stalingrad area were Gumrak and Pitomnik. It is late December 1942, and a Ju 52/3m transport of KGr.zbV.700 is shovelled clear of the overnight snow. At that time the daily quota for the Luftwaffe transport pilots was two supply flights to Pitomnik in the 'cauldron'. This picture was taken at Tazinskaja, one of the support airfields.*

Bottom: *In December 1942, Soviet tank spearheads reached the landing ground at Simovniki which had to be evacuated by the Luftwaffe. Here the* Staffelkapitän *of* 4./St.G.77 Oblt *Alex Gläser (centre) and the* Gruppe *Medical Officer (right) are waiting for a Ju 52/3m transport scheduled to take them and the remaining personnel to safety. In the background, a single-engined Junkers W 34 courier aircraft.*

Top: *Despite the confused situation Ju 88A bombers of III/KG 3 based at Kharkov attempt to give some relief to the remaining troops of the Sixth Army in the ruins of Stalingrad.*

Centre: *The Luftwaffe machines parked on the snow-covered airfield at Dniepropetrovsk look almost like model aircraft – the FW 189, He 111, Fi 156 and Ju 52/3m Despite many newly built hangars most of the Luftwaffe aircraft had to remain parked in the open.*

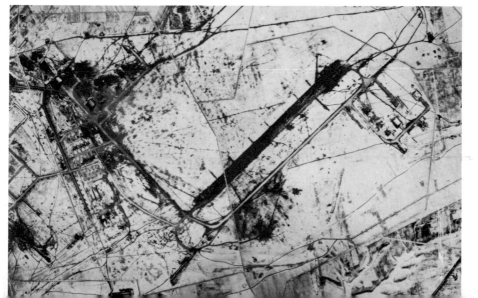

Bottom: *A bird's eye view of the airfield at Orel in January 1943. The Luftwaffe formations based at Orel supported the battles of the Army Group 'Centre' during its advance and subsequent retreat.*

Top left: *All who had 'a rank and a name' in II/JG 3 'Udet' in June 1943 are seen together in this picture. Standing from right are Ofw Eberhard von Boremski, Ofw Georg Schentke and Fw Siegfried Engfer. Sitting in front of them is Oblt Viktor Bauer.*

Bottom left: *Fw Otto Kittel was the most successful fighter pilot of JG 54 'Grünherz' when he was killed in action over Courland in Latvia on 14 February 1945. His score stood at 267 confirmed aerial victories.*

Top right: *Oblt Karl Schrepfer (centre), Staffelkapitän of 6./St.G.1, shortly before another operational flight from Volchansk on 30 June 1942. In October 1944 Schrepfer was appointed commander of III/SG 1.*

Bottom right: *A day to remember! Oblt Hans-Ulrich Rudel is greeted by his men on returning from his 1000th operational flight at Gorlovka on 10 February 1943.*

Top: *Just like in the West, the Bf 109 soon found itself tasked with forced high-speed reconnaissance in the East as well. Here is the T5+TL, a Bf 109F–5 of 3.(F)/Aufkl.Gr.Ob.d.L. at Luga in March 1942.*

Centre: *Only a few examples of the Arado Ar 240A were built, one of which, the GL+QB, was being temporarily used by 3.(F)/100 long-range reconnaissance Staffel. However, the pilots could not get used to the flight behaviour of this odd aircraft. The machine is depicted at Kharkov in February 1943.*

Bottom: *This captured Soviet aero sledge served in a more sporty role with IV/JG 51 'Mölders'. The odd-looking craft was powered by a light-aircraft engine driving a pusher propeller. It was an original Soviet design, widely used on the Eastern Front. Fitted with armour plating, these aero sledges spearheaded attacks across snowy expanses; others were used as ambulances or for liaison.*

Top: *The air war in the Polar regions in winter 1942/43 had some light-hearted moments as well. Some of the more prominent figures of JG 5 are seen here 'at ease' in front of one of their aircraft. From left:* Fw Hans Döbrich, Lt Theo Weissenberger, Oblt *Heinrich Ehrler (Staffelkapitän),* Ofw. *Rudolf Müller and* Ofw. *Albert Brunner – all from 6./JG 5.*

Bottom left: *The Bf 109G–5 fighters of JG 5 'Eismeergeschwader' Taxiiing to their take-off runway at Alakurti in Northern Finland. Apart from escort tasks the JG 5 pilots were also engaged on fighter sweeps and low-level attacks, often far behind the Soviet lines.*

Bottom right: *After re-equipping with the FW 190A, two non-commissioned pilots of I/JG 54 pose for the war correspondent's camera in front of their new mount,* Ofw *Fritz Tegtmeier (left) and* Ofw *Hans-Joachim Kroschinski. Krasnogvardeisk, January 1943.*

Top: *On 13 June 1943 III/KG 55 celebrated its 10,000th operational flight at Stalino. Known as 'Greifengeschwader', KG 55 remained equipped with twin-engined He 111 bombers until the end of the war.*

Centre: Lt *Walter Krupinski, known to his comrades of 6./JG 52 as 'Graf (Count) Punski', climbing out of his 'brown 5', a Bf 109G–3, after an operational flight.* [*Krupinski became one of the highest-scoring Luftwaffe fighter pilots of the war with 197 confirmed victories. He ended the war flying Me 262 jets.* Tr.]

Bottom: *The first Luftwaffe night 'Behelfskampfstaffeln (auxiliary bomber squadrons) were formed in October 1942 as direct counterparts of the increasingly troublesome Soviet U–2 and R–5 night nuisance bombers. These units were renamed 'Störkampfstaffeln' in November 1942, and became* Nachtschlacht-gruppen (NSGr) *in October 1943. By late summer 1944 there were 11 such NS-Gruppen on the Eastern Front including two flown by Latvian and Estonian volunteers and one Staffel by ex-Soviet Air Force pilots. They flew mostly converted training aircraft such as the Gotha 145 shown here.*

Top: *In addition to the I1-2* Shturmoviks *the Soviet Air Force also used the Sukhoi Su-2 as low-level close-support aircraft, which appeared in 1943. The picture shows a Su-2 forced down by German fighters near Roslavl. The Su-2 was ordered into production in 1940 and known then under the functional designation system as BB-1 (close-range bomber). Just over 100 were in service in June 1941, by which time designation had been changed to Su-2. They sufferd a very high attrition rate and production stopped in 1942.*

Bottom left: *Beginning 1943, the allied Royal Romanian Air Force received increasingly modern German aircraft. A Ju 87D of a Romanian dive-bomber squadron at Jassi in May 1943.*

Bottom right: *A Soviet pilot who had crashed with his damaged aircraft into a water tower near Minsk is being treated by German medical personnel. A Luftwaffe staff doctor/Hptm stitches lacerations in the Soviet lieutenant's face.*

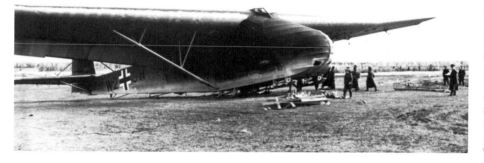

Top: *Deserting or captured Soviet Air Force pilots could volunteer to fly with the 1.Ostfliegerstaffel d.Luftwaffe. Equipped with captured Soviet aircraft carrying German markings this unit operated on the Eastern Front for several months in 1943/44.*

Centre: *The Me 321 Gigant (Giant), a large-capacity cargo glider, could carry a useful load of up to 20 tons, equal to the total carried by eight Ju 52/3m transports. During the early period of its service use the Me 321s were towed to their destination by three twin-engined Bf 110 fighters in so-called 'Troika-schlepp' (Troika Tow). The Me 321 – and the unsuccessful Ju 322 Mamut – were originally designed for the planned airborne invasion of England.*

Bottom: *The Me 321 towing problem was solved by the five-engined He 111Z, a combination of two He 111H airframes connected by a new wing centre section fitted with the fifth engine. From then on, the towing Gruppen on the Eastern Front could tackle this heavy glider with ease.*

Top: Hptm *Erich Hartmann, the Richthofen of the Second World War. His record score of 352 confirmed aerial victories stands unbeaten to this day. In this picture* Ltn Erich 'Bubi' Hartmann *of 9./JG 52 appears quite pleased with his modest 121 victories to date marked on the rudder of his Bf 109G. Novo Zaporozhe, 2 October 1943.*

Centre: *A Luftwaffe brass band performing its regular Sunday concert in a small Ukrainian village.*

Bottom: *A German field cemetery near Krivoi Rog. All such burial places were marked on maps before retreat, but can no longer be found. On higher orders, all German field cemetaries were deliberately ploughed over within weeks by the victorious Soviet forces.*

Top: *The 2000th operational sortie of 1.(F)/124 was flown on 29 June 1943 by the Ju 88D with Oblt Spitz's crew. The 'Jubilee' crew after landing at Kirkenes, Norway, with (from left) Ofw Reissig (radio operator), Oblt Spitz (observer), Ofw Schön- bucher (pilot) and Fw Pohl (gunner).*

Centre and Bottom: *To support their guerrilla operations the Soviets established regular landing fields behind the German lines which were used to fly in agents, weapons, radio sets and supplies. It was the task of special units to find and capture such guerrilla air bases and this modified U–2 (Po–2) fell into the hands of one such detachment. The two large plywood containers fitted under the wings could each accommodate two fully armed guerrillas.*

165

Top: *'Tank busters' of
13.(Pz)/SG 9 in front of one
of their Hs 129B aircraft at
Hranovka in February 1944.
At that time the tank
destroyer group was led by
Hptm Hanschke (2nd from
left), while the 13.Staffel had
just been taken over by
Hptm Oswald (left). The two
pilots of the squadron
sharing the joke are Fw
Kotter and Ofw Heger.*

Centre: *The Crimea peninsula had to be abandoned
by the German forces in April/May 1944. During
this evacuation operation heavy equipment was
transported across the Black Sea by the so-called
'Siebel ferries' and landed in Romanian ports.
Whenever possible, these moves were protected by
Luftwaffe fighter escorts.*

Bottom: *During the
evacuation of Minsk in June
1944 a small group of
Luftwaffe signals auxiliaries
were left behind, completely
to their own devices. At the
last moment, the girls
managed to find shelter
among the photographic
personnel of Aufkl.Gr.100,
already loaded on a train.*

166

Top: *The commander of SG 2, Hans-Ulrich Rudel, flying his fully armed Ju 87G 'tank destroyer' on another operational sortie. By the end of the war Rudel's personal score amounted to 519 Soviet armoured fighting vehicles destroyed, for which he was awarded the Golden Oakleaves to the Knight's Cross.*

Centre: *The* Staffelkapitän *of 10.(Pz)/SG 2, Lt Anton Korol, destroyed 99 Soviet tanks and was awarded the Knight's Cross at Kamennz in March 1945. Both he and his gunner/radio-operator Fw Karl Müller recorded a total of over 660 operational flights each.*

Bottom: *By summer 1944 the majority of Stuka-* Gruppen *had been re-equipped with the FW 190F close-support aircraft. A FW 190F–3 of II/SG 77 taking off on an operational sortie from Lemberg in summer 1944.*

167

Top: *Until Finland had to cease fighting against the Soviets on 4 September 1944, the Finnish fighter squadrons flew their Bf 109G and Fiat G.50s side by side with the Luftwaffe against the common enemy over the Karelian Front. A Finnish Fiat G.50 fighter at Rovaniemi in April 1944.*

Centre: *Luftwaffe bomber formations managed a final big punch on the night of 21 June 1941 when they struck the American 'shuttle bombing' base at Poltava, destroying 47 four-engined bombers and 15 fighters on the ground. This German aerial reconnaissance photograph shows the parked B–17s at Poltava.*

Bottom: *The 3./JG 52, known as the 'Caramba-Staffel' experienced the final days of the war led by their* Staffelkapitän Oblt *Erwin Günther at Deutsch-Brod. Here the men are carrying him shoulder-high back to the barracks after his latest aerial victory. A few days later the fighting was over for these young pilots.*

9. North Africa and the German Retreat from the Mediterranean

After the occupation of Greece by Axis powers on Mussolini's wishes the administration of the country was transferred to the Italians in June 1941. The German military command limited its activities to the extension of some ports as naval and air-sea rescue bases and the improvement of a number of airfields for use as advanced bases by flying formations. However, most of the *Staffeln* subordinated to X *Fliegerkorps* were operating from North African soil at that time.

These units comprised I/St.G 1 and II/St.G 2 under the *Geschwaderstab* of St.G 3, which had supported the advance of the DAK (German Africa Corps) to the Egyptian border, bombed the fortifications and port of Tobruk during the siege (1 June–5 December 1941), and then covered the withdrawal battles from December 1941 onwards. The fighter protection of the two *Stukagruppen* was in the hands of I/JG 27 together with 7./JG 26, while the heavy fighters of III/ZG 26 escorted supply ships and transport aircraft across the Mediterranean. Additional support in fighting at the Egyptian front was available in the shape of Ju 88A bombers of LG 1 and He 111Hs of II/KG 26 stationed on their well-constructed airfields in Greece and on Crete.

The fact that X *Fliegerkorps* was totally occupied with the interests of the DAK during the battles in Cyrenaica had given the British garrison of Malta a most welcome break; it was used by British convoys to bring in supplies and additional armament, which soon made this island base into a serious interference factor in Axis supply convoys and the air transport flights, still sparse, to North Africa. The intolerable loss rate of transport ships and tankers in November 1941 dictated an urgent resumption of aerial attacks on Malta and an early run-up to Operation 'Herkules', the planned attack on and occupation of the island by Axis forces. However, all this was hopelessly beyond the capabilities of the Italian *Regia Aeronautica* and for that reason bomber groups had to be found elsewhere – pulled back from the Luftwaffe units deployed on the Eastern Front.

The formations commanded by the C-in-C South, GFM Kesselring, landed on their new airfields in Sicily in December 1941. According to Luftwaffe strength returns on the Eastern Front this transfer left a major gap in Luftflotte 2, which had provided most of the formations.

With the bomber *Gruppen* I/KG 54, II/KG 77, III/KG 77, KGr. 606 and KGr. 806, all equipped with Ju 88As, a total of 150 bombers was now available for the task. This Malta force was reinforced by I/NJG 2 equipped with Ju 88Cs and III/St.G 3 with Ju 87D dive bombers. Fighter escort for these raids, which lasted until April 1942, was provided by all three *Gruppen* of JG 53 and II/JG 3 with their Bf 109Fs.

Almost cut off from supplies, with only a few serviceable aircraft left in the underground hangars at the end of April 1942, the garrison of Malta would have been almost helpless to resist Operation 'Herkules'. Preparations for it had already been started by the commanding general of XI *Fliegerkorps*. According to plan, German and Italian parachute troops were to be set down on the island, supported by a surprise attack of about 300 cargo/assault gliders landing another 3000 troops, while a transport fleet would land six Italian divisions at various points on the island. All of a sudden, Hitler had second thoughts and stopped everything. He mistrusted the Italians, had doubts about their fighting spirit and, apart from that, saw no real need to face the risk involved in this combined invasion after General Rommel's successful spring offensive in North Africa. All available transport and glider towing aircraft were ordered back to their parent formations and the invasion of Malta postponed indefinitely.

From the initial positions at Marada/Mersa-el-Brega, Rommel, with his German–Italian

Panzerarmee, had started a new offensive on 21 January 1942, aimed at retaking the lost Cyrenaica territory. Exactly five months later, on 21 June 1942, it was crowned with the capture of Tobruk with its enormous stocks of supplies. The advance was carried by its momentum across the Egyptian border north of the Quatara Depression, right up to El Alamein. There the British had constructed deeply laid defensive positions on which the Axis offensive ran aground late in June 1942 – not least because of enormously extended supply lines and all the difficulties that this situation entailed.

As an offensive force during the Axis advance and the following position warfare *Fliegerführer Afrika*, Gen Lt Hoffmann von Waldau had the entire St.G 3, 12./LG 1 and the *'Jabogruppe Afrika'* (fighter-bombers), while JG 27, III./JG 53 and 10./ZG 26 countered the increasing number of intrusions by the British Desert Air Force, which was gaining in strength all the time. The sole short-range reconnaissance *Staffel*, 4.(H)/13, equipped with the Hs 126 and Bf 110s, provided battlefield reconnaissance in close cooperation with the DAK, while for longer-range reconnaissance tasks Luftflotte 2 had 1.(F)/121, 2.(F)/122 and 2.(F)/123. A rather special unit was the 1.*Wüstennotstaffel* which, with its specially equipped Fi 156 Storchs, carried out search-and-rescue sorties over the desert.

Within LF 2, the formations of II.*Fliegerkorps* were based in Sicily (Stab and II/JG 53, I/JG 77, Stab, I, II and III/KG 54, Stab and I/KG 77, while those of X.*Fliegerkorps* were divided between Greece and Crete – *Jagd-Kommando* 27, III/ZG 26, Stab, I and II/LG 1, 6./KG 26, II and III/KG 77, II and III/KG 100). It was evident that there had been some changes since the spring.

Not surprising, but unexpected in its extent and force, came the British counter-offensive at El Alamein, beginning on the night of 23/24 October 1942. Under Lt Gen Montgomery, the British Eighth Army forced its way forward and within four months succeeded in driving back the German–Italian forces all the way through Marmarica, Cyrenaica and Tripolitania right up to the Tunisian border. During this time the aerial superiority lay clearly with the Desert Air Force, which was constantly supplied with the latest aircraft types in generous quantities. Things were quite different on the German side: the *Staffeln* under *Fliegerführer Afrika* were already beginning to feel flying equipment shortages and the onset of supply difficulties. Bringing up reserves, equipment, fuel and ammunition caused the most headaches for the German command.

The Axis having failed to take Malta at the time when it had been practically bled white, several convoys from Alexandria had reached the island in June, and during the following months Malta was supplied with all that was necessary for it to defend itself again. The replenished stock of aircraft in particular soon made Malta into a dangerous 'wasps' nest'. From then on, the losses suffered by II *Fliegerkorps* began to mount quite noticeably and during the late summer of 1942 the bombers could only attack Malta when escorted by a strong force of fighters. By the end of the year even that no longer paid dividends and the Luftwaffe bombers began increasingly to raid Malta at night.

Together with the fighter squadrons of the Desert Air Force whose bases had moved westwards with Montgomery's advance, the Malta-based squadrons now began to take an increasingly heavy toll of the transport aircraft plying between southern Italy and North Africa, which eventually led to the catastrophic situation in spring 1943. Under *Lufttransportführer Mittelmeer* (air-transport leader Mediterranean), Gen Maj Buchholz, the available transport *Gruppen* (including the KG.z.b.V. I/323 equipped with six-engined Me 323s) were formed into two large formations in November 1942 – the *Geschwader* 'N' (Naples) and 'S' (Sicily). In wide, loosely deployed formations of up to 100 aircraft, protected by just one or at the most two *Staffeln* of fighters, the Ju 52/3m, Me 323, Fiat G.12 and Savoia-Marchetti SM.82 transports would fly across the Mediterranean towards their landing bases at Tripoli and Tunis (from December 1942, only Tunis and Bizerta). Heavy losses suffered while approaching the coastal area and attacks by Allied fighter-bombers during unloading so weakened both *Geschwader* that by December 1942 they were already down to about two-thirds of their establishment. They were also ordered to hand over some aircraft for the Stalingrad airlift, and by the end of the year only about 200 Ju 52/3m and 20 Me 323 transports were left to cope with supplying the Axis forces in North Africa. It was only in March 1943 that these formations could be to some extent filled up again.

The landing of American and British forces in Morocco and Algiers (Operation Torch) on 7/8 November 1942 represented a threat from behind to the DAK, then involved in tough defensive battles, and additional danger to the landing fields of the transport formations. Soon afterwards, the loss rate of the transport units under *LT-Führer Mittelmeer* showed a sudden steep increase. The following examples depict this trend quite clearly: 10 April 1943 –

19 aircraft lost; 11 April – 18 lost; 18 April – 24 lost; 22 April – 16 lost (including 14 six-engined Me 323s carrying fuel). The fighter escort, always inferior in numbers, was helplessly exposed to swarms of British and American fighters at this phase. No assistance could be expected from the French squadrons in North Africa; the Vichy French had ceased all resistance against the invading Allied troops on 12 November 1942. The reaction of the German government was swift: immediate occupation of Vichy France (Operation Anton) and the island of Corsica.

Pressed hard from two sides, the surrounded 252,000 Axis troops of the *Heeresgruppe Tunis* (Army Group Tunisia) surrendered on 13 May 1943. Only a few members of the DAK managed to escape to the Italian mainland, or to be flown out. With that, fighting in North Africa came to an end.

Crowded hopelessly on the defensive both on the Eastern Front and in the Mediterranean the Wehrmacht was forced in mid-1943 to prepare for the defence of the so-called '*Festung Europa*' (Fortress Europe). The loss of North Africa threatened an invasion of the Italian mainland in the near future and called for the necessary preparations. All aircraft remaining serviceable after covering the final battles in Tunisia had been flown across to Sicily, Pantelleria or southern Italy and after a short period of rest and replenishment the units had moved into their newly allocated airfields. When the Allied invasion of Sicily began on 10 July 1943, it set off in motion the planned German countermeasures under the keyword '*Fall Marder*' (Case Pine Marten). At that moment LF 2 had seven fighter *Gruppen*, two *Gruppen* of SKG 10 highspeed bombers, 14 bomber *Gruppen*, one close support/assault *Gruppe*, two (H)-*Staffeln*, three (F)-*Staffeln* and a fluctuating number of transport and naval-aviation *Staffeln*.

The landing of the American Seventh Army at Gela and Licata on the southern shores, and of troops of the British Eighth Army at Avola and Syracuse further eastwards, were preceded by devastating area bombing raids on Axis airfields in Sicily as well as a bombardment of the actual landing zones by heavy naval units. The few German fighter *Staffeln* of JG 53 and JG 77 remaining on Sicily tried to cover the desperate attacks on the Allied beachheads by Luftwaffe close support/assault and bomber *Gruppen*, but had to retreat to the Italian mainland on 13 July 1943. A day later, He 111s of TGr.30 dropped Luftwaffe paratroopers on the Catania airfield, but despite bitter resistance Sicily had to be given up on 17 August 1943. During these defensive battles the Luftwaffe had constantly attacked various land and particularly naval targets, paying a high price in men and machines.

On the night of 25 July 1943, after a meeting of the High Council of the Italian fascist party had declared no-confidence in Mussolini and Marshal Badoglio was asked by the Italian king to form a new government, came the '*Fall Achse*' (Case Axis). This meant immediate disarmament of Italian armed forces, including the navy and the air force, but was not in fact implemented until 8 September 1943 after the Italian High Command had signed a secret separate armistice behind the back of its German ally. Right in the middle of these disarmament measures burst the news of the landing of strong Anglo-American forces at Salerno, which necessitated the formation of a blocking position across the Italian 'boot'. The German formations still at Apulia and Calabria were pulled back behind this new front.

As the German ground forces were constantly forced to retreat northwards, so the Luftwaffe formations also had to move their bases rearwards. But it was a hard time. The Luftwaffe airfields were attacked daily by four-engined bombers and fighter-bombers, and the operational units found it increasingly difficult to fulfil their tasks, combatting the Allied assembly positions and supply routes, and particularly carrying out surveillance and constriction of the bridgeheads. The loss of Foggia was especially serious because its capture late in September gave the Allies a broad airfield country ideal for medium and heavy bombers; the latter could now extend their range right up to former Czechoslovakia.

But there were also setbacks for the Allies. On 1 August 1943 a B–24 Liberator formation took off from Benghasi in North Africa to attack the Ploesti oil wells in Romania, then only just within their range. However, due to an error by the lead navigator this raid cost the US Ninth Air Force 53 heavy bombers and 310 aircrew. Apart from everything else, the long approach flight over the Mediterranean, Albania and Macedonia precluded any fighter escort. All that changed with the capture of the airfields around Foggia, San Severo and Bari; from then on such targets were within range of the twin-boom P–38 Lightning fighters and enabled them to escort bombers raiding targets in Greece, Yugoslavia and Romania.

On the Luftwaffe side, bomber *Gruppen* of LG 1, KG 1, 6, 30, 54, 76 and 77 had remained in Italy and from their bases at Perugia, Grosseto, Piacenza, Frosinone and Viterbo had a relatively short approach flight to their targets. Other units, such as KG 26 and KG 100 were based at Montpellier, Salon-de-Provence and Istres in southern France; their targets were the Allied convoys in the western Mediterranean and the Allied bridgeheads in Italy. The two close-support/ground-assault *Gruppen* had to be based near the front lines so that they could reach their targets carrying the optimum 1000kg (2205lb) bomb load. As for the fighters, they could operate from any reasonably large meadow; the determining factor was availability of well-camouflaged parking spaces.

The sinking of the fleeing Italian battleship *Roma* on 9 September 1943 and the liberation of the 'Duce' from his enforced exile on the Gran Sasso were notable events in which the Luftwaffe played the leading role. A new Allied landing in the Nettuno-Anzio area on 22 January 1944 made the Italian capital Rome – just 50km from this beachhead – a frontline town. Fierce and sacrificial attacks by the Luftwaffe ground-support and bomber *Gruppen* (Operation *'Morgenröte'* = Dawn) could not prevent the expansion of the landing zone. The taking of Rome by the Anglo-American forces on 4 June 1944 led to the final phase of fighting in Italy. By January 1945 the front had stabilized along the line Lucca-Firenzuola-Ravenna in northern Italy.

The attempt by the Germans, after the fall of Mussolini, to build up an Italian air force that would be an effective ally largely failed, partly due to inadequate co-operation and the emotional nature of the situation. However, some Italian airmen did remain on the German side, and formed the Air Force of the Italian Socialist Republic, which fought on until the end of hostilities in Italy. They were supplied with the latest German Bf 109G fighters and other equipment, as well as flying the latest Italian designed fighters. Beginning mid-September 1943, a large part of the *Regia Aeronautica* aircraft park had been transported to Germany. Many of the transport aircraft flew with Italian crews; some aircraft, with German markings, were incorporated with Luftwaffe training formations; others were used by operational Luftwaffe formations. The Italian transport aircraft produced by Savoia-Marchetti (SM.82) and Fiat (G.12) continued to be built under German supervision in their parent works for several months afterwards; work was also continued under RLM auspices on several projects at the Caproni and Piaggio plants.

After the break-up of her alliance with Italy, Germany had to avoid a situation in which Greece and the Aegean islands, then under Italian administration, would be surrendered to the Allies without a fight. For that reason, the disarmament of Italian forces there had to take place as swiftly as possible. Italian island garrisons which resisted these measures came under determined attacks by Luftwaffe units from their bases in Greece to soften them up for the following occupation by German forces: Corfu and Rhodes in September and Leros in September–November 1943.

In contrast with the scattered partisan groups operating in Greece, the bandit activities in Yugoslavia under its leaders Mikhailovic and Josif Broz-Tito had taken forms that called for uncompromising action by all available German forces, including the Luftwaffe. Fortunately the feared Allied landings along the Albanian and Yugoslav coasts had not taken place; they would have resulted in a quick and total collapse of the whole situation in the Balkans. Some of the Italian garrison troops in Yugoslavia had deserted to the guerillas, taking with them their armament, including artillery. Actions against these guerilla armies, such as *'Maibaum'* (Maypole), *'Waldrausch'* (Forest ecstasy) or *'Rösselsprung'* (Knight's move – in chess) often took place in cooperation with German paratroops and Luftwaffe units. The Luftwaffe flying formations in the Balkans, including training and replacement units such as SG 151 and parts of NSGr 4 and 7, attacked identified bandit camps and positions by day and by night. It was a different, brutal kind of war: a forced landing or bailing out over the bandit area meant certain death. An example for many was the fate of the fighter pilot *Hptm* Joachim Kirschner of IV/JG 27, bearer of the Oak Leaves to Knight's Cross, who was summarily executed by the bandits on 17 December 1943. German officers who ordered countermeasures against these 'guerillas' who, according to the Geneva Convention, were not soldiers but illegally armed civilians, were accused of war crimes after the war and sentenced to death or long-term imprisonment. Two well-known examples were *Oberst* Bruno Bräuer, commander of Fj Rgt 3 (Parachute Rgt) and *Generaloberst* Alexander Löhr, commander of LF 4 and later C-in-C of Army Group E, who were sentenced to death and executed in Greece and Yugoslavia respectively.

Ever since the Allied landings in Italy, the Luftwaffe bases at Tatoi, Eleusis, Argos and Larissa in Greece were exposed to constant air raids and bombed until no longer tenable. In any case, the fighter *Gruppen* stationed in the Balkan area (parts of JG 4, 5, 27, 51, 53 and 77) were never strong enough to cope with the escorted formations of Allied four-engined bombers intruding first from North Africa and later from southern Italy; their own losses had already risen way above all expectations. With the collapse of Romania on 23 August 1944 and the defection of Bulgaria three days later the whole eastern flank of the Balkans had become completely exposed, and Hitler ordered withdrawal from the Aegean. Evacuation of the islands and bringing back of personnel and material remained the tasks of the German Navy as well as of the naval-aviation and Luftwaffe transport formations. These actions succeeded without undue losses, as did the previous evacuation of Corsica and Sardinia (in September/October 1943).

Despite the fact that the Allies had complete aerial superiority over the Balkans, the Soviet and Bulgarian forces attacking from the East managed to cut off only parts of the German Army Group E retreating from Serbia (late in September 1944), forcing these units into the area controlled by Montenegrian bandits. Operations over the narrow 'pipeline', the only remaining way out for the retreating German troops, were the last large-scale offensive sorties undertaken by the Luftwaffe in the Balkans, ending in November 1944.

The Soviet Union signed armistice agreements with Bulgaria on 8 September 1944 and Hungary on 20 February 1945, thus turning both countries into open enemy territory. As a result, the frontlines had moved dangerously close to German territory in the south-east as well.

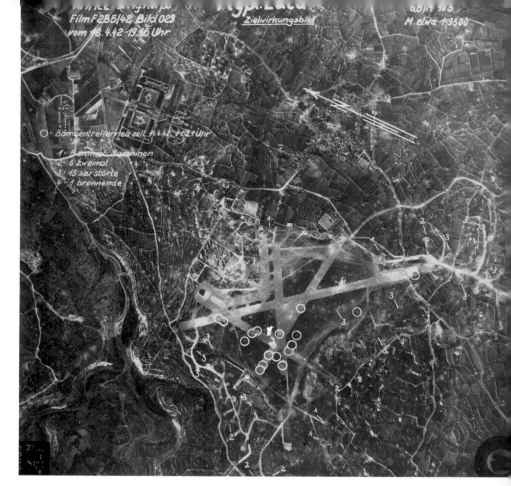

Top: *To carry out the planned occupation of the Mediterranean fortress of Malta it was essential to achieve absolute aerial superiority over the island. To that end, the airfields of La Venezia and Luca were among the principal targets of the bomber formations in spring 1942.*

Centre: *An unusual view of a spit of land at the port of La Valetta as seen over the starboard engine of a Ju 88A of KGr.806. Clearly visible are the barrack buildings and an anti-aircraft gun position. The main port of Malta too was the daily target of the Luftwaffe and Italian bombers.*

Bottom: *After Hitler had abandoned the planned attack on Malta in April 1942, the deployed towing formations returned back to the flying training schools in Germany. A line-up of DFS 230A assault/cargo gliders behind a Do 17 towing aircraft.*

174

Top left: *A Ju 87 B–2 of II/St.G.1 moments before commencing its dive on targets in the port of Tobruk. At the time this photograph was taken, November 1941, the garrison of Tobruk had been surrounded for five months and still repulsed all attacks by the Afrika Korps.*

Top right: *The acquisition of target information for the bomber formations in the Mediterranean area was the task of 2.(F)/123. On 23 August 1942 this reconnaissance squadron flew its 2000th operational sortie, with the 'Jubilee' crew seen here after landing back at the base. From left: Uffz Merz (radio operator), Gefr Fastenrodt (gunner), Ofw Nitsch (pilot) and Oblt Bournot (observer).*

Bottom: *To achieve the optimum fragmentation effect the 50kg (110lb) SD 50 splinter bombs were fitted with special 'stand-off pipes', the so-called 'Dinort-Spargeln' (Dinort asparagus). These bombs exploded just above the ground and were especially effective against groups of vehicles and troop assembly areas. They were named after their inventor, Oskar Dinort, a noted Stuka pilot and commander.*

Top: *The crew of* Hptm *Joachim Helbig made a special name for themselves as 'ship killers' in the Mediterranean. Pictured in the cabin of their Ju 88A are (right) Helbig (then* Staffelkapitän *of 4./LG 1) with his radio operator* Ofw *Franz Schlund standing behind him.*

Centre: *A Ju 88A–4 of II/KG 77, camouflaged to blend with its surrounding terrain, at Gerbini in October 1942. In front of it (right) is* Hptm *Paepcke, commander of the* Gruppe.

Bottom: *A successful crew of 2.(F)/122 is visited by the commander of their* Gruppe *at Trapani in October 1942. From left:* Oblt *Schiedel (pilot),* Oberst *Fritz Koehler (*Gruppe *commander),* Ofw *Berlacher (gunner) and* Ofw *Hartwig (radio operator). Fitted with auxiliary fuel tanks the Ju 88 reconnaissance aircraft could reach far into Egypt.*

Top left: *During the battle for Solum the crew of* Hptm *Hermann Hogeback were especially successful. From left: Hogeback (Staffelkapitän of 9./LG 1 in August 1941), Fw Lehnert (radio operator) and Uffz Reschke (observer). The gunner, Uffz Glasner, missed this ride on a bomb. Some time later* Hptm *Hogeback was awarded the Oak Leaves with Swords to the Knight's Cross for his leadership and achievements.*

Top right: *During their night operations against the British in Cyrenaica in February 1942 the He 111Hs of KG 26, the famed 'Löwengeschwader', wore black underpaint. Here is Lt Herbert Kuntz, a pilot in II/KG 26, with the mascot of the* Gruppe.

Bottom: *No information has survived as to whether production by this 'Egg Factory' of the German Luftwaffe in North Africa managed to satisfy the daily demand. Recorded at Berka in North Africa, in June 1942.*

Top left: *At the time this picture was taken, in February 1942, Lt Hans-Joachim Marseille of 3./JG7 had already proved, with his 48 confirmed victories, that he had overcome his initial difficulties. Within the following seven short months he was to add another 110 enemy aircraft to his score. [With 158 confirmed aerial victories, Marseille was the highest-scoring German fighter pilot in combat against Western Allies. His record series of victories in one day – 17! – was never equalled. The date was 1 September 1942, and all 17 aircraft he claimed shot down have been confirmed from Allied records. Tr.]*

Top right: *A detachment of the Afrika Korps fires the last salute over the open grave of Hptm Marseille. The young master of aerial combat lost his life when the engine of his new Bf 109G fighter caught fire on 30 September 1942; while bailing out Marseille was apparently knocked unconscious by the tailplane and fell to his death in the desert below.*

Bottom: *In addition to the Fi 156 Storch the Luftwaffe desert emergency Staffel also had some twin-engined FW 58C Weihe aircraft for search-and-rescue tasks. A Weihe is shown here wearing the typical Luftwaffe North African desert camouflage: olive-green patches on sandy-brown background.*

Top: *British supply ships damaged by Axis bombers in the port of Tobruk. Its garrison withstood the long siege in 1941 and Tobruk was not taken until 21 June 1942, during the second Rommel offensive.*

Centre: *The severe losses of German and Italian shipping led to a change in tactics and from early 1942 on, increasing use was made of air supply of Axis forces in North Africa. Shown here is a Go 242 cargo glider in tow behind a Ju 52/3m. [The main reason for these shipping losses was information obtained via ULTRA at Bletchley Park: by mid-1941 the British were able to read most of the secret Luftwaffe 'Enigma' machine-code messages dealing with supplies and the deployment of units. Tr.]*

Bottom: *Flying in formation, protected by only a few fighters, the Luftwaffe transport aircraft had to suffer heavy losses at the hands of attacking Beaufighter, Spitfire and Kittyhawk fighters. A group of Ju 52/3m transports of II/KG.zbV.1 on the route Malemes/Crete to Derna/North Africa in June 1942.*

179

Top: *During the later stages the Luftwaffe transport formations in the Mediterranean area were reinforced by the six-engined Me 323 large-capacity transports. But the cost was high: in April 1943 II/TG 5 lost almost 75 per cent of its establishment of Me 323s. Its size alone made the lumbering Me 323 an easy target for RAF fighters. Note the then novel 'clamshell' loading doors and the tail-down position of the aircraft, resting on the rear two pairs of its 10-wheel undercarriage – both features were introduced by the Me 323 and later adapted elsewhere.*

Centre: *As the twin-engined He 59 floatplanes had proved unsuitable for ASR service under tropical conditions the all-metal Do 24 flying boats became the preferred aircraft for that task. A Do 24T taking off on a search flight from Taormina in Sicily.*

Bottom: *In 1942 the three-engined BV 138C–1 flying boats of 1.(F)/SAGr.125 were used for maritime reconnaissance and submarine search tasks over the Aegean Sea and the Mediterranean in general. Armed with two heavy and one light machine gun, the aircraft possessed sufficient defensive fire power to look after itself.*

180

Top: Generalfeldmarschall *Kesselring, commander of Luftflotte 2, in conversation with* Fliegerführer Afrika, Gen Lt *Otto Hoffmann v. Waldau. Due to the spread of the Mediterranean theatre of war it was necessary to have constant formal agreements and coordination of all Luftwaffe air operations.*

Bottom left: *A pilot from a transport* Staffel *being assisted to strap on his back parachute before the next operational flight. To be able to move freely inside an aircraft, the observers, radio operators,*

Top right: *The commanding general of* Luftflotten-Kommando Süd-Ost *(South-East)*, General der Flieger *Martin Fiebig, during an inspection flight with a He 111 in December 1943.*

gunner and other flying personnel were equipped with clip-on parachutes (bottom right). During the flight the parachute pack was naturally kept within easy reach.

Top: *While based at Bucharest-Pipera in July 1941 III/JG 52 received its first brand-new Bf 109F–4 fighter which is seen here receiving appropriate looks of astonishment. Just a month later the* Gruppe *was transferred to the Eastern Front, flying operationally in the Lower Dniepr region.*

Centre: *These He 112B fighters of the Royal Romanian Air Force stemmed from German deliveries. As the He 112 had lost out to the Bf 109 during the comparison trials in autumn 1936 it was not generally accepted for service in the Luftwaffe, although twelve He 112B fighters, already sold to Japan, were repainted and impressed into Luftwaffe service during the Sudeten crisis in 1938, equipping part of III/JG 132 until the signature of the Munich Agreement. A total of 24 He 112B fighters were sold to Romania prewar and delivered between May and September 1939.*

Bottom: *The IAR–80 was a modern aircraft of Romanian design which began to join the Royal Romanian Air Force fighter squadrons early in 1942. The IAR–80 was used both for local defence and on the Eastern Front.*

Top: Oblt *Emil Omert, Staffelkapitän of 8./JG 77 at Arco Philaenorum during the British Eighth Army counteroffensive. The III/JG 77 had arrived in North Africa late in October 1942.*

Centre: *The 9./JG 53 'Pik As' (Ace of Spades) based at Trapani in Sicily was one of the units detailed to escort transport formations on their way across the Mediterranean to North Africa. The Staffelkapitän Hptm Franz Götz with two NCO pilots of his squadron in April 1943.*

Bottom: *The only night fighter unit in the Mediterranean area was I/NJG 2 which had been deployed there in November 1941. Equipped with Ju 88Cs, this Gruppe was operational on convoy escort duties, carried out raids on Malta, and made long-range night-fighting sorties over Egypt and daytime attacks as the occasion demanded. Its principal base was at Catania.*

Top: *Until April 1943 all available aircraft of the Luftwaffe Air Transport Commander Mediterranean were required for supply flights to Africa. Afterwards, the same squadrons – partly equipped with Ju 52/3m floatplanes – had to take care of the evacuation from the surrounded Tunisian battle-zone.*

Centre: *The crews of LTS See 222 (Air Transport Staffel 222–Sea) with their BV 222 Wiking large-capacity flying boats were also ordered to join the evacuation of casualties from North Africa. The BV 222V–5 X4+EH at the footbridge of its moorings at Piraeus in Greece. Originally designed as a commercial flying boat in 1937–38, only 13 BV 222s were completed. The X4+EH was later destroyed by RAF fighter-bombers while moored at Biscarosa in France in June 1943.*

Bottom: *The low-level attack by a large formation of B–24 Liberator bombers on the Ploesti oil refineries in Romania cost the American Ninth Air Force dear. A B–24 burning out on the ground after an attempted crash landing. Of the 179 B–24s despatched 165 attacked, not all of them bombing Ploesti and 43 B–24s were lost over the target. Another eight more-or-less damaged B–24s sought refuge in neutral Turkey, and only 114 B–24s returned to base, some to crash-land.*

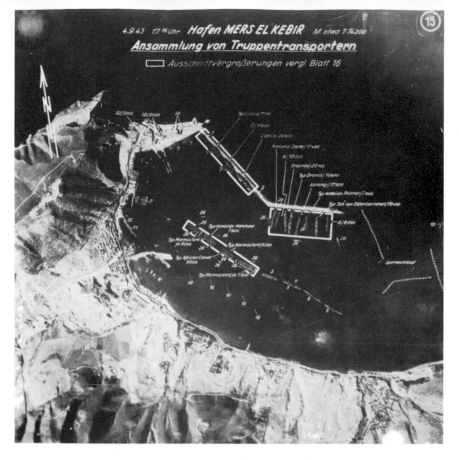

Top: *An aerial reconnaissance photograph taken by Luftwaffe aircraft on 4 September 1943 nearly three months after the invasion of Sicily points to another operation. It clearly shows the Allied transport fleet assembled at Mers el Kebir taking on board troops and war material. This new landing followed just five days later, at Salerno.*

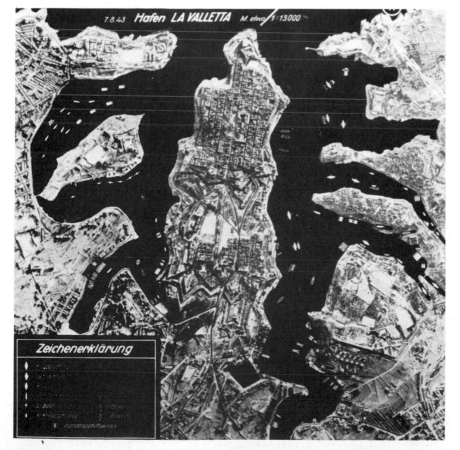

Bottom: *After the Allied jump over the Mediterranean the strategic significance of Malta became even more important. This Luftwaffe aerial reconnaissance photograph of La Valetta, taken on 7 August 1943, shows the port teeming with transport and landing vessels.*

185

Top: *The ubiquitous Ju 52/3m also found another use. Fitted with a large, energised light-metal loop, the so-called 'Mausi-Ring', it would be regularly employed to clear waters endangered by Allied electro-magnetic mines, such as entrances to ports and river estuaries. These aircraft belonged to the three mine clearance squadrons operated by the Luftwaffe.*

Centre: *To carry the optimum offensive load against the Allied bridgeheads at Salerno, Anzio and Nettuno the FW 190s of the Italian-based close-support formations would take one 1000kg (2205lb) SC 1000 bomb each. Because of its length and size, one fin had to be sawn off for ground clearance at take-off.*

Bottom: *After the deposed Duce had been liberated by German airborne commando troops led by Otto Skorzeny on 12 September 1943 a Fi 156 Storch was instrumental in flying out the main party, Mussolini and Skorzeny. Although the available rock-free space at Gran Sasso was far too small, the pilot* Hptm *Gerlach just managed to get airborne. The original intention was to fly out Mussolini alone – who was also a good pilot himself – but Skorzeny apparently insisted on coming along and his 6ft plus frame of solid bone and muscle nearly proved too much under the circumstances.*

Top: *A Ju 88-Staffel of LG 1 returns to its base in Greece after an operational sortie against the Allied invasion troops in Southern Italy on 24 October 1943.*

Bottom: *To get away from pursuing fighters, Ofw Bach of 1.(F)/123 'pressed' his Ju 88A–6/U so low down that its propellers touched the Mediterranean. Despite the missing tips the pilot managed to bring his aircraft safely back to its base at Perugia. Several such cases are known, clear proof of the wisdom of fitting the Ju 88 with the special Schwarz-type wooden propellers. This 'splintered end' effect also avoided 'bending' the engine crankshaft, a far more dangerous and expensive damage.*

Top left: *By August 1943 the aircraft of KG 26 had sunk 32 vessels in operations against Allied supply shipping in the Mediterranean. At that time this formation, which had already proved its worth in similar strikes in the North Sea, was commanded by* Major *Klümper.*

Top right: *The Luftwaffe torpedo-bomber crews were trained at Grosseto in Italy. Two training torpedoes are being attached to the under-fuselage carriers of an He 111H–6.*

Bottom: *The crews of III/KG 77 achieved some good successes in attacks against American convoys sailing off the Algerian coast. After taking off from Istres in southern France, the Ju 88A–17 3Z+UT carrying two torpedoes is flying towards the target at low altitude. Note the typical wave-pattern camouflage used by many Luftwaffe torpedo-bombers.*

Top left: *After the Badoglio government had changed sides the Italian troops at Leros surrendered the Dodecanese islands to the Allies without a fight. Before their reoccupation by German airborne troops the Luftwaffe carried out several bombing raids on Allied positions and barracks in October–November 1943.*

Top right: *Reinforced by Italian deserters, the activities of Yugoslav guerilla bands increased considerably from autumn 1943 onwards. An operational squadron of SG 151, a close-support training unit, is photographed on a bombing sortie against identified guerilla nests in the Montenegro mountains on 22 October 1943.*

Bottom: *The disarmament of Italian troops on Syros took place without incident. The date is 15 September 1943, and an Ar 196A of 1./SAGr 126 has just taxied into the port of Ermoupolis.*

Top: *After landing paratroops and cargo gliders on Leros on 12 November 1943, Ju 52/3m floatplanes arrive with additional weapons and supplies and fly out the first casualties.*

Centre: *A Ju 88T–1 of 2.(F)/123 has just landed and a motorcycle combination is immediately on hand to take the valuable film cassette to the photographic centre. Important decisions often depended on speedy evaluation and forwarding of reconnaissance results.*

Bottom: *Reconnaissance of Allied supply ports and convoys as well as surveillance of the many islands was hard work for Luftwaffe long-range reconnaissance crews. Even after the introduction of the improved Ju 188D the squadrons still suffered losses at the hands of Allied fighters.*

Top: *On operational trials, the Ju 252 pre-production machines transported bulky cargoes to and from the battlefields in the East and the Balkans. A hydraulically operated tail ramp facilitated quick and convenient loading and unloading of the aircraft. The Ju 252, incidentally, was intended as replacement for the seemingly eternal but limited capacity Ju 52/3m. The small pre-production series built proved their worth, but due to the raw material situation it was decided to abandon this all-metal transport for a much simpler mixed-construction redesign, the Ju 352 Herkules, which was built in series.*

Centre: *A number of Me 323 Gigants of TG 5 and Ju 88Cs of I/NJG 2 fell victim to an Allied 'bomb carpet' on Castelvetrano in Sicily on 13 April 1943. Such massed air raids on airfields in Sicily and the Italian mainland prepared for the Allied invasion of Southern Europe.*

Bottom: *After gaining possession of several air bases in southern Italy the American bombers turned their attention to Luftwaffe airfields in Greece. Burning hangars following an area bombing raid on Saloniki airfield.*

191

Top: *After the Stalingrad catastrophe the activities of foreign air forces allied to Germany began to show a distinct reduction. These ex-Polish PZL P–37A and P–37B bombers of the Royal Romanian Air Force carried out only a few operational sorties on the Eastern Front early in 1943. Here they are seen parked in the open at Kronstadt in Romania in winter 1943/44.*

Centre: *The Bulgarian fighter squadrons were equipped with some Bf 109Es supplied by Germany at the beginning of the Second World War. Together with the Bf 109G fighters delivered from 1943 onwards these served for territorial defence purposes.*

Bottom: *The Croat armed forces supported the Luftwaffe on the Eastern Front with one bomber and one fighter squadron. Two Croat Breguet 19 B2 biplanes with different engines are seen here at a training school in Sarajevo in spring 1943. The Croat volunteer bomber squadron served as part of III/KG 3 from Oct 1941 to early 1942, and then again as 15.(kroat.)/KG 53 from July 1942 to Nov 1942.*

Top: *An airborne action against the headquarters of the Yugoslav guerilla chief Tito took place near the small town of Drvar, south-west of Banja Luka, on 25 May 1944. Known under the covername* 'Rösselsprung' *(Knight's move), it involved German air-landing troops and the sole parachute battalion of the Waffen-SS. A DFS 230A glider of LLG 1 (air-landing Geschwader 1) is seen in the foreground.*

Bottom left: *General Alexander Löhr (centre) commanded Luftflotte 4 during the Balkan campaign. He was later accused of war crimes, sentenced to death by a Yugoslav court-martial and executed.*

Bottom right: *Warned in advance, Josef Broz-Tito managed to get away and all that* 'Rösselsprung' *captured in his headquarters were two British liaison officers and Tito's Marshal's uniform.*

Top: *A He 46 night-attack bomber of NSGr.4 after taking off from Mostar on its way to bomb guerilla hide-outs in July 1943. Visible under the fuselage is the special 'grate' for the small SD 2 splinter bombs.*

Bottom: *An American 'bomb carpet' dropped on Easter Sunday 1944 destroyed the largest part of the Romanian aircraft factory at Kronstadt (today Brasov). A number of Romanian IAR–80 and German Bf 109 fighters are parked in the foreground.*

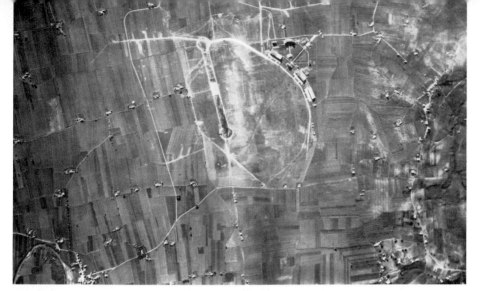

Top: *In the course of the continuing withdrawing movement the Ju 88s and Ju 188s of 1.(F)/123 found a new refuge at Perugia, north-east of Rome. An aerial view clearly shows the airfield installations outside the town.*

Centre: *In spring 1944 a number of former* Reggia Aeronautica *aircraft were repainted with Luftwaffe markings and flown to Germany. This picture shows a Caproni Ca 313G navigation trainer accompanied by a He 111 of the Ferrying Detachment which was subordinated to the General of Flying Training.*

Bottom: *Adjusting the weapons of a Reggiane Re 2002 fighter on a firing range. Hardly any of the Italian fighters saw service with operational Luftwaffe formations; they were mainly transferred to fighter-trainer schools.*

Top: *One of the few Luftwaffe fighter* Gruppen *in the south-eastern theatre was II/JG 51. Here the 'white 7' of Lt Elias Kühlein of 4./JG 51 is waiting for the next operational sortie at Nis in Yugoslavia.*

Centre: *Slightly wounded in the face by a sliver from his shattered cabin glazing, Lt Günther Stedfeld (Staffelkapitän of 4./JG 51) takes it easy together with Lt Kühlein (Staffelkapitän of 5./JG 51) at Komorn in Hungary in October 1944.*

Bottom: *Most of the Italian Savoia-Marchetti SM.82 Cangaru aircraft were taken over by the Luftwaffe transport formations. The SM.82 had a much more spacious cargo room than the Ju 52/3m and was also faster. It remained in Luftwaffe service until the end of hostilities. The SM.82 in the picture seems to have 'bent' its port undercart.*

196

10. War in the air over Germany and the occupied Western regions

After the postponement of Operation *'Seelöwe'* (Sealion) and the shifting of the theatre of war to south-eastern Europe in April 1941 only Luftflotten 2 and 3 remained in occupied France, Belgium and Holland, comprising the following formations:

Luftflotte 2:	1.(F)/122	Ju 88
	2.(F)/122	Ju 88
	3.(F)/122	Ju 88
	4.(F)/122	Ju 88
	I/SKG 210	Bf 110, Me 210
	I and II/KG 3	Ju 88A
	I, II and III/KG 4	He 111H
	I/KG 28	He 111H
	I, II and III/KG 30	Ju 88A
	I, II and III/KG 53	He 111H
	I, II and III/JG 51	Bf 109F
	I and II/JG 52	Bf 109E
	I, II and III/JG 53	Bf 109F
Luftflotte 3:	1.(F)/123	Ju 88, Bf 110
	II and III/KG 1	Ju 88A
	I, II and III/KG 27	He 111H
	I and II/KG 54	Ju 88A
	I, II and III/KG 55	He 111H
	I, II and III/KG 76	Ju 88A
	I, II and III/KG 77	Ju 88A
	KGr.100	He 111H
	KGr.606	Ju 88A
	KGr.806	Ju 88A
	I, II and III/JG 2	Bf 109E, 109F
	I, II and III/JG 26	Bf 109E, 109F

Fliegerführer Atlantik (attached to LF 3):

	3.(F)/123	Ju 88
	I/KG 40	Fw 200C
	II/KG 40	Do 217E
	III/KG 40	He 111H

General d.Luftwaffe at Ob.d.M. (High Command, German Navy): all coastal aviation, catapult and air-sea rescue formations

Luftwaffen-Befehlshaber Mitte:

	I/JG 3	Bf 109F
	I and II/JG 54	Bf 109F
	Erg.St./ZG 76	Bf 110 (reserve/ replacement unit)
	I, II and III/NJG 1	Bf 110D
	I/NJG 2	Ju 88C, Do 17Z
	I/NJG 3	Bf 110D

The post of *Luftwaffen-Befehlshaber Mitte* (Luftwaffe Commander Centre) had been established in March 1941 and embraced all those formations that did not come under any Luftflotten but were deployed for the defence of Germany. It was only logical that in January 1944 this post and its subordinated formations would be renamed Luftflotte Reich.

With the main effort concentrated on nocturnal raids on the British Isles the Luftwaffe fighter *Gruppen* were largely released from escort tasks and mainly had to combat the increasing number of RAF intrusions. In addition to the already tactically proven long-range night-fighting (I/NJG 2) there were also the beginnings of successful interception night-fighting, whereby searchlights coupled to ground radar sets would lay hold of enemy aircraft and lead the night fighter orbiting in the near-by 'waiting space' towards it. Once the night fighter pilot had spotted the enemy aircraft illuminated by searchlights, he would attack visually. This was known as the *'Helle Nachtjagd'* or *'Henaja'* (Bright night-fighting).

The abundance of bomber *Gruppen* in Western Europe in April 1941 shrank quickly after the start of the Eastern campaign. The actual *'England-Geschwader'*, KG 2 and KG 6, did not move into their bases in the West until 1942, although one of them, KG 2, had already participated in raids on the British Isles in 1940; in June/July 1941 it was on the Eastern Front, still equipped with the dated Do 17Z, but was then withdrawn from operations and retrained on the new Do 217E at Achmer beginning that autumn. First to become operational under LF 3 in the West was II/KG 2 in

February 1942. On the other hand, KG 6 was a completely new unit, formed in autumn 1942 from various contingents and sections furnished by other bomber *Geschwader*.

Convoy reconnaissance and attacks on shipping targets were the principal tasks of *Fliegerführer Atlantik* created in March 1941. Appointed to this post was *Obstlt* Harlinghausen whose brief was to cover the Atlantic and especially the Gibraltar route with the aircraft under his command, operating from bases around Bordeaux. Initially still flying the He 111 and Do 217 bombers, the three *Gruppen* of KG 40 were gradually re-equipped with the Fw 200C and He 177 for longer-range operations.

The task facing the day- and night-fighter formations during Operation 'Cerberus' (11–13 February 1942) was not of the everyday kind. It was a matter of providing air cover for the two German battleships *Scharnhorst* and *Gneisenau* and the heavy cruiser *Prinz Eugen* sailing from Brest along the Channel to their new bases in Norway. The operation was carried out by concentrating all available forces (about 250 day and night fighters and 'heavy fighters') directly subordinated to *General der Jagdflieger, Oberst* Adolf Galland. While the bomber and reconnaissance *Staffeln* commanded by General Coeler kept the enemy naval forces off the backs of the German naval formation, the day and night fighters repulsed attacks by more than 200 British naval aircraft and, together with the naval anti-aircraft guns, shot down 49 attackers. The Luftwaffe losses in contrast totalled only nine aircraft.

The month of February 1942 was also notable for the British Commando raid on a German ground radar site near Bruneval on Cap d' Antifer. The British Operation 'Bite' took place on the night of 27/28 February with the drop of 120 parachute troops who, with minimal own losses, managed to remove parts of the FuMG 62 'Würzburg' radar and take them back to England in waiting motor boats. In fact, the paras held the perimeter while a couple of RAF technicians who had volunteered for this dangerous job managed to dismantle some parts of the Würzburg. This operation exemplified the determined efforts of both sides to gain predominance in the radar technology race.

The introduction of the Mosquito night fighters and new air interception radar in the RAF were reflected in a marked increase in Luftwaffe losses during night raids on England from mid-1942 onwards despite the fact that the participating bomber formations were equipped exclusively with the faster and more agile Ju 88 and Do 217s by that time. The He 111H had already completely disappeared from combat formations in the West.

The Luftwaffe fighters too had improved their standing: re-equipment with the more modern Fw 190A had begun in late July 1941 (II/JG 26). The Fw 190A was powered by a portly air-cooled radial engine which was far less susceptible to combat damage than the liquid-cooled power plant of the Bf 109. Nevertheless, the fighter-bomber (*Jabo*)-*Staffeln* remained operational with their Bf 109Es and Fs well into 1942. The high-altitude interception *Staffeln*, 11./JG 2 and 11./JG 26, were still flying the more powerful Bf 109G-version in 1944.

Other formations, such as JG 5, JG 52 and JG 53 remained 'pure' Bf 109-*Geschwader* till the end of the war and were always supplied with the latest production series to replace their 'windmills' which would be worn out after a short period of time.

By late 1941 the single-engined Bf 109 and twin-engined Bf 110, designed from the outset as pure fighter aircraft, were also being introduced by the long-range reconnaissance *Staffeln* operating over England. Fitted with additional fuel tanks and with reduced armament, these fast aircraft were ideal for the 'forced reconnaissance' sorties which had become a necessity over southern England. However, when it comes to a medley of unusual aircraft types, the reconnaissance *Staffeln* of Ob.d.L. (C-in-C Luftwaffe) with their mixture of such machines as the Ha 142, Ar 240, Ju 86P, Do 215 and various pre-production aircraft, had a place of their own. Compared to that, the Bf 109s and Bf 110s, together with the Ju 88s, seemed quite 'ordinary' as reconnaissance aircraft.

As with aircraft types, constant development was also taking place in the technical fields of offensive loads and aircraft armament. Although the drop of the first SC 2500 'Max' (2500kg/5511lb) large-capacity bomb on London on 21 December 1940 had made headlines, the trend had changed more towards drop containers filled with smaller bombs. Released at higher altitudes, these containers would disintegrate at a predetermined height and scatter their load over a larger area. 'Area bombing', as first demonstrated by the RAF during its heavy night raids of March 1941, had now also become the practice of the Luftwaffe bombers.

The aircraft armament too had undergone a change. The MG 15 as flexible and MG 17 as fixed weapons, both of 7.92mm calibre, had by mid-1942 been largely replaced by the 13.1mm MG 131, 15.1mm MG 151 and 20mm MG 151/20.

Together with an increase in calibre the thicker armour protection had also made it necessary to develop improved ammunition with better penetrating and explosive powers.

As a result of the German–Italian declaration of war on the USA on 11 December 1941, the Americans had started preparations in February 1942 for the formation of the US Eighth Air Force in England. The first aircraft and crews arrived in Britain in July 1942, and the new air force opened its activities on 17 August with a daytime raid by 12 B–17 Fortresses, accompanied by an escort of four squadrons of RAF Spitfires, on railway installations at Rouen in German-occupied France. The B–17s gave the rather hesitant attacking German fighters their first impressions of a close-order formation of four-engined bombers and their defensive firepower. Only two of the Fortresses were damaged – by anti-aircraft guns. The next American raid on 6 September 1942 was flown by 41 B–17s, the target being the Potez works at Meaulte in France, but this time they lost one B–17 shot down and brought back several damaged aircraft. The German fighters had begun to show their teeth.

If the plans of the US Air Force in England comprised bombing raids on military and strategically important targets in Germany and German-occupied territories from the very beginning, the Britain's Air Marshal Harris, commander of the RAF Bomber Command since February 1942, declared his determination to continue unrestricted bombing warfare against Germany.

The high loss rate suffered by RAF bombers during daytime intrusions at the beginning of the war had acted as a catalyst on British air warfare strategy and it was decided to achieve the goal, the destruction of the German armament industry, by means of nocturnal bombing raids, in the knowledge that 'wide bombs' would hit dwelling houses. Understandably, night raids with then still unreliable bombsights did not offer the same accuracy as bombing in broad daylight. Now however, when the responsible German authorities had also declared the distinction between industrial centres and dwelling settlements as 'insignificant', the destruction of German towns had been openly accepted as the RAF bombing programme. It began with massed raids on Lübeck on the night of 28/29 March 1942; Rostock on 24/25 April 1942; and then Cologne on 30/31 May 1942.

With this intensification of the bombing in large formations, the night fighting tactics hitherto used by the Luftwaffe were no longer suitable. 'Bright night fighting' (*Henaja*) could combat only one enemy aircraft 'captured' by searchlights, while in 'dark night fighting' (*Dunaja*) a night fighter would be guided by a ground radar set to an enemy bomber detected by another ground radar set and then 'talked' into position by a fighter guidance officer on the ground – a complicated, time-consuming and sometimes useless method. The only solution was to concentrate on perfecting the so-called '*Freie Nachtjagd*' ('free night fighting') in which night fighters carried their own search devices. In this case, a 'Würzburg-Riese' (Giant Würzburg) ground radar only had to find the enemy bomber and continually inform the fighter guidance officer regarding its position. Their task was then to 'talk' the night fighters orbiting in the nearby 'waiting spaces' into the bomber formation. Once they had been 'sluiced' into the bomber stream, the night fighter crews were on their own and had to make the best use of their wide-angle air-interception radar sets (FuG 202 Lichtenstein BC: search angle 70°; later FuG 220 Lichtenstein SN–2; azimuth 120°; elevation 100°). Once an enemy bomber had been detected, the night fighter pilot would be guided by his radar operator sitting behind him until visual contact when it was again up to the pilot to choose the best attack position and open fire.

Beginning May/June 1943, in addition to the fixed armament concentrated in the fuselage nose and ventral gun pannier, the Bf 110, Do 217 and Ju 88 night fighters received another weapon system. At first fitted individually by operational maintenance and repair workshops, and later installed on the production line, this comprised obliquely upwards-firing sets of guns built into the cockpit or in the fuselage just behind it, the so-called '*Schräge Musik*' ('Slanted Music'). With this novel armament, once he had visual contact, a night fighter pilot could position himself underneath the enemy aircraft avoiding the defensive weapons in the tail and dorsal turrets. Flying in parallel under his target, the pilot would then use his upwards-pointing reflex gunsight to aim at the enemy bomber and fire his oblique guns. This method of combating the nocturnal bombers helped a number of the leading Luftwaffe night fighter pilots to achieve their impressive victory scores. [This deadly method was 'invented' almost simultaneously by several German and Japanese night fighter pilots. An experienced pilot would always aim for the wing sections between the two inner engines, where the big fuel tanks were. The whole idea was to destroy the bomber, giving its crew a chance to bale out as well. Only a fool would fire into the fuselage; if the bombs were still inside, the resulting explosion would rip both aircraft apart! The amazing thing is that reports, by returning RAF crews, of 'vertical fire followed by explosion of a bomber' were ignored by RAF command for over a year! *Tr.*]

The ground and air interception radars had their own problems. Apart from technical faults, still frequent in these early days, enemy jamming such as the already familiar staniol strips (known as 'chaff' by the Allies and '*Düppel*' by the Luftwaffe) would occasionally cause serious interference, but special attachments or alternative frequencies would quickly make the 'blind see again'.

Interim and makeshift solutions such as the inclusion of single-engined fighters in night fighting, although initially remarkably successful, did not work on a long-term basis. The machines were basically day fighters, fitted with insufficient instrumentation and, despite the proven blind-flying experience of their pilots, the losses soon began to mount. Another drawback was the need for special optical expedients such as directional searchlights, the so-called 'light streets' in the night sky, orientation or control points fixed by several coned searchlight beams, and direction-indication anti-aircraft fire using special ammunition. This '*Wilde Sau*' (Wild Boar) night fighting also created confusion in proven night-fighting zones established under the aegis of *General* Kammhuber, Inspector of Night Fighters. For these reasons, early in 1944 the 'Wild Boar' JG 300, 301 and 302, formed after the first Hamburg 'firestorm' night of 24/25 July 1943, were transferred to day fighters lock, stock and barrel.

As long as the heavy American bombers, operating at their maximum combat radius, could only be protected by limited fighter escort which left the bombers to cover the final part of their route to the target on their own, the Luftwaffe dayfighters were able to book considerable defensive successes. Thus, during a raid on the Fieseler aircraft works at Kassel and the AGO aircraft plant at Oschersleben on 23 July 1943, the Americans lost 22 of the 77 B–17s involved, because their fighter escort had to turn back with half-empty tanks at the German border. However, the biggest reverse of this 'bloody summer' (as the American airmen called it) came during their double raid on Schweinfurt and the Messerschmitt plant at Regensburg on 17 August 1943. From the very moment the escort fighters turned back at the Dutch border, the formation of 370 four-engined American bombers was 'crowded' by concentrated attacks by Luftwaffe single- and multi-seat day-fighter and night-fighter *Gruppen* throughout the approach and return flight, being shot at with

guns of all calibres, including the newly introduced 210mm Wfr.Gr.21 rocket shells. The Bf 109 and Fw 190 fighters could carry one rocket launcher tube, the Bf 110 a pair of launching tubes under each wing. These rocket shells were fired at 'closed' bomber formations from beyond the range of their defensive guns, the idea being to 'blow apart' the tight bomber 'box' and thus scatter its concentrated defensive fire before attacking individual bombers. This tactic was successful as long as the bombers were not accompanied by escort fighters; with their underwing mounts the Luftwaffe fighters were very hard pressed to hold their own against their more agile opponents.

On 14 October 1943 the Americans tried another large-scale raid by unescorted bombers, once again aiming at the German ball-bearing centre at Schweinfurt. This time the price paid was even higher: no less than 64 B–17s fell to the German defenders and another 130 reached their British bases in badly shot-up condition. [According to US Eighth Air Force records 291 B–17s sent out, 229 were effective (i.e. reached the target area); 60 B–17s went missing in action, 7 were written off in crashes in the UK. 138 landed with battle damage. B–17 crews claimed no less than 186 Luftwaffe fighters shot down, 27 'probables' and 89 damaged – more than the Luftwaffe had in action on that day! Actual Luftwaffe losses: 35 shot down, 20 damaged. *Tr.*]

Fitted with drop tanks beginning June 1943, the American P–47 Thunderbolt escort fighters were able to accompany their charges about 20–30km deeper into Germany – which was not really much of an improvement on their range. Then came the P–51 Mustang, the first of which arrived in England late in 1943. The Mustang proved a first-rate long-range escort fighter and soon the B–17 and B–24 formations could have reliable fighter cover all the way to their targets and back.

The so-called 'Big Week', a planned series of raids on the German aviation industry, began on 20 February 1944. The P–51 Mustang escort fighters hung like bunches of grapes above and beside the bomber 'boxes' and parried the attempts by Luftwaffe fighters to get through to the 'fat autos' (Luftwaffe slang for US four-engined bombers). The air war over Germany had entered a new phase.

The Luftwaffe night raids on England had begun afresh in March 1943. *Obstlt* Peltz, an experienced bomber pilot and commander, had been appointed *Angriffsführer England* (Attack leader England) and, with concentrated efforts of KG 2 and KG 6, assisted by the newly formed pathfinder unit I/KG 66 (established in June 1943), had managed to unload 2320 tons of bombs on the British Isles by the end of 1943 – an insignificant fraction of the 136,000 tons of bombs dropped on Germany and German-occupied territory by the RAF alone in that year.

With Operation 'Steinbock' (Ibex), a series of heavy 'revenge' raids ordered by Göring, it was now intended to pay back the British in the same coin for their deliberate nocturnal terror raids on German towns. To this end, 'Angriffsführer England' was allocated bomber formations from other Luftflotten, his force eventually consisting of replenished KG 2, KG 6 and KG 54, parts of KG 30, KG 40, KG 66, KG 76 and KG 100 as well as I/SKG 10 (Fw 190 fighter-bombers). Operation 'Steinbock' lasted from 21 January until late May 1944 when the very heavy loss rate of 60 per cent (about 300 aircraft) made the continuation of this purely prestige action unjustifiable. For that, the southern part of the British Isles were now to come under the hail of the first 'revenge weapons', the V1 (also known under the cover designations of FZG 76 = Flakzielgerät 76, or Anti-aircraft target, and 'Kirschkern' = Cherry stone) which, after a long period of test firing, was now practically ready for operational use.

All plans regarding the future forms of German aerial warfare had to be completely reassessed after the Allied landing on the European Continent on 6 June 1944. For a start Allied aerial superiority made it urgently necessary to stop all offensive aerial warfare in the West. It was essential now for the Germans to protect their own skin, so to speak, and combat the daily intruding bomber streams with such accuracy and effectiveness that the absolute British–American aerial mastery could be gradually reduced and perhaps even completely broken. In other words, the fighter output had to be increased far above its present rate; such were now the 'needs of the hour'.

Until that time, the figure of 1050 fighters produced in July 1943 (725 Bf 109s, 325 Fw 190s) had been the highest monthly total; the output figures for single-engined fighters during the following months were also around the 1000 mark. However, the concentrated air raids on the German aviation industry in February 1944 had led to a drop in aircraft output – in some plants up to 75 per cent – and this shortfall now had to be made good using immediate emergency measures. A special Jägerstab (Fighter staff) was set up on 1 March 1944 under the overall direction of Generalluftzeugmeister (Chief of Aircraft Procurement)

Erhard Milch and the Minister of Armaments, Speer, and subordinated to Speer's deputy, Otto Saur. Given special authority, this new command instigated the quickest possible improvement and repair of the bomb-damaged plants, provision of additional labour force and, above all, immediate decentralization and dispersal of production facilities, in some cases into disused mines or other tunnels.

The success of this Jägerstab is shown by the output figures: over 2000 aircraft in April and 2200 in May, with production reaching its all-time high in September 1944 with no less than 3375 fighter aircraft.

In addition to standard operational types this production programme also included various new developments, such as the Ta 152 high-altitude interceptor, Ju 388 high-altitude reconnaissance bomber, He 219 night fighter and Me 262 jet fighter (known under the covername of 'Silber' = Silver). By that time the priority attached to the aircraft known as 'Zinn' (Pewter), the Me 163B Komet, had already been downgraded; the flight endurance and range of this rocket-powered object defence interceptor did not accord with the new thinking.

The ferrying of most of the newly completed aircraft from their assembly plants to the operational units was the responsibility of Flugzeugüberführungsgeschwader 1 (aircraft ferry Geschwader), divided into Gruppen North, East, South, South-West, West and Centre. This formation, abbreviated as Fl.Ü.G.1, had been established on 20 May 1942 out of the aircraft-ferry department of the Luftwaffe supply office. In addition to Luftwaffe pilots no longer fit for operational service and works pilots, its personnel also included female 'drivers' who, provided they had the necessary qualifications and abilities, could volunteer for service in this formation. Despite the non-operational character of the task, ferrying of unarmed machines to the 'aircraft sluices', distribution centres or directly to the operational units was no holiday during the months of Allied aerial supremacy. Official casualty reports of Fl.Ü.G.1, with names which carry an added note 'Due to enemy action', are a clear proof of this.

The high production figures of the German aviation industry allowed not only regular material replacement of losses suffered by operational formations but also the creation of a fighter reserve which could be made use of in case of further production shortfalls. According to Adolf Galland, then General of Fighters, on 12 November 1944 the Luftwaffe had 3700 aircraft and pilots ready for action in

18 fighter *Geschwader*. His intention was to use this force to attack a large-scale American air raid, and by shooting down as many aircraft as possible weaken the enemy both materially and psychologically. However, an order from higher authorities held back the fighter reserve for an important ground-support operation and so it never came to the hoped-for 'mopping up' among the four-engined bombers.

The use of fighters in ground support during the ill-fated Ardennes offensive beginning 16 December 1944 and Operation *'Bodenplatte'* (Baseplate), a surprise strike on Allied airfields during the early hours of 1 January 1945, broke the back of the Luftwaffe day fighter force. Prepared in the strictest secrecy, Operation *'Bodenplatte'* was a Pyrrhic victory costing 151 killed or missing Luftwaffe fighter pilots, plus another 53 captured – a loss rate the weakened fighter *Geschwader* could no longer recover from. True enough, some 465 British and American aircraft were left wrecked or damaged on their airfields, but the Allies suffered hardly any personnel losses.

Perhaps the most tragic aspect of this operation was the fact that due to secrecy hardly any German anti-aircraft batteries were informed about the forthcoming attack and opened fire on their own aircraft (any large formation of aircraft was automatically assumed to be Allied at that time!), shooting down or forcing to crash-land no less than 90–100 Luftwaffe fighters!

After this latest bloodletting the American and British bombers and fighter-bombers roamed across the remaining German territory almost unhindered. The replacement Luftwaffe fighter pilots, by then ever younger and less experienced, did not survive on average longer than four operational sorties. This, together with the acute fuel shortage, did not allow more than the occasional pin-prick against the Allied air armada.

The initiation of self-sacrifice attacks following, albeit imprecisely, the Japanese 'Kamikaze' pattern, in this phase of the war, already hopelessly lost, shows the 'end of the world' mood that befogged the minds of some of the German leadership at that time, but which to some extent affected everybody who experienced those days. A formation of such volunteers was known as *'Sonderkommando Elbe'* (Special Detachment Elbe) who were assembled and waiting for their literally last sortie in the Gardelegen-Stendal area. The pilots of the detachment were supposed to close right in to the four-engined American bombers before opening fire and, when out of ammunition, to ram them! On 7 April 1945 a still not precisely known number (100–180) of Bf 109s and Fw 190s of *'Sonderkommando Elbe'* took off and, escorted by some Me 262 jet fighters, set course for an American bomber formation. Contact was established north of the Mittelland canal over Central Germany. In separate sections, the 'Elbe' pilots dived through the American fighter escort towards the B–17s and B–24s, in some cases effecting a suicide attack. Contradictory reports mention 133 German losses as against 50 American, but even if these figures were reversed, such a recourse would have been completely meaningless in terms of the final outcome of the war. After all, by that time the front lines in the West already ran along the right bank of the Rhine, and in the East – through Pomerania and just before the gates of Berlin.

The Me 262-equipped *Gruppen* of JG 7, KG(J) 54 and JV 44 also occupied a doomed outpost. It was certainly a unique feeling for a pilot who flew these aircraft to be faster than the enemy, but that feeling would change into depression when, during take-off and landing, their jet fighters, unable to make the slightest defensive movement, were 'plucked' by the roaming Tempests, P–51 Mustangs and P–47 Thunderbolts.

The fact that the night fighters still managed to inflict painful losses on the enemy during the final months of hostilities was due in part to the experienced cadre aircrews who had to face only one enemy at a time until the end (although sometimes they also had an Allied long-range night fighter sitting on their necks), and in part to the RAF aircrews themselves who had grown careless out of a feeling of superiority. The 'wall of fire' that faced the night fighter was never as thick as that the day fighters had to go through to come to grips with the bombers.

The scene offered by the Luftwaffe airfields, aircraft plants and accessory works at the end of the war was one of sheer desolation. The heaps of wrecked aircraft in destroyed hangars, on bombed hardstandings, in 'splinter boxes', on 'shadow airfields' and other dispersals did not look anything like the former 'Sword in the sky', which now had to sacrifice itself in being forced to follow a megalomaniac leadership into the abyss.

Top: *While the majority of fighter formations were deployed East for Operation Barbarossa, JG 2 'Richthofen' remained in the occupied Western areas. Pilots of 8./JG 2 discuss an escort task at Le Havre in October 1941.*

Centre: *Individual fighter-bomber squadrons formed within fighter Geschwader carried out high-speed daytime raids over Southern England in 1942. That summer 10.(Jabo)/JG 26 'Schlageter' had its Bf 109F–4/B fighter-bombers housed in well-camouflaged board structures at Caen-Carpiquet.*

Bottom: *Fighter escort during Operation Cerberus (the 'Channel dash' of the major German Navy units from Brest to Germany) during the period 11–13 February 1942: Bf 109s patrol over the heavy cruiser* Prinz Eugen *to protect her from enemy torpedo and bomber attacks.*

Top: *The Luftwaffe operational bases in Northern France sometimes offered quite remarkable sights. The hangars, workshops and accommodation buildings were 'changed' to fit the surrounding countryside by painting on doors and walls. Others went even further: II/KG 27 'Boelcke' based at Lille-Nord let the harvest go right into the 'barn'.*

Centre: *Another hangar at Lille-Nord was camouflaged with mock roofs and painted-on windows and doors to make it look like a large farmhouse.*

Bottom: *The artists at Vitry-en-Artois excelled themselves by 'changing' the hangar doors into a whole street section with a suburban cafe and a hairdressing salon. However, once the positions of the Luftwaffe airfields were known to the Allies such camouflage efforts became rather ineffective.*

Top left: Hptm *Joachim Müncheberg (left) commanded II/JG 26 'Schlageter', part of the so-called 'Kanalgeschwader' in spring 1942. Sitting next to him is* Oblt *Wilhelm-Ferdinand Galland, a brother of the former* Geschwader-Kommodore. Hptm *Müncheberg was a most accurate shot and achieved a total of 135 confirmed aerial victories. He lost his life over Tunisia on 23 March 1943 when a wing broke off his aircraft in a combat turn.*

Top right: *On 15 August 1941* Reichsmarschall *Göring arrived at the Buc airfield near Versailles to watch a demonstration of new aircraft types. Behind him is his host,* Major *Kopper, commander of Aufkl.Gr.123, the 1.Staffel of which was based at Buc.*

Bottom: *The JG 2 'Richthofen' was commanded by Major Walter Oesau from July 1941 to June 1943. He is seen here with pilots of his staff flight (Stabsschwarm). From left:* Oblt *Rudolf Pflanz,* Oblt *Erich Leie,* Major *Oesau and* Lt *Egon Mayer. [All were exceptional fighter pilots. Rudolf Pflanz achieved 52 confirmed victories in the West before being shot down by RAF Spitfires over France on 31 July 1942; Erich Leie scored 118 victories (43 in the West) before losing his life in a collision with a Yak–9 on the Eastern Front on 7 March 1945; Walter Oesau had 125 victories to his credit (73 West, incl. 10 4-engined bombers) when he was shot down by P–38s on 11 May 1944; and Egon Mayer had achieved 102 victories (all West, including 25 four-engined bombers) when he was killed in a dogfight against several P–47s on 2 March 1944. Tr.]*

Top left: *The 'electric fire' at Bruneval on Cap d' Antifer. After the unknown object had been identified on aerial photographs for what it was, a British commando unit landed by parachute on 27 February 1942 succeeded in removing important parts of this FuG 62 radar set and take them back to England.*

Top left: *The first Do 217-Js were delivered to Luftwaffe night fighter formations in the West early in 1942. They were armed with 4×20mm MG FF cannon and 4×7.92mm MG 17 machine guns in the fuselage nose, an exceptionally powerful forward-firing armament. Their martial look was intensified by the FuG 212 (Lichtenstein C–1) radar aerial array.*

Bottom: *The first experimental installation of a FuG 212 (Lichtenstein C–1) air interception radar in a Do 17Z–10 Kauz II. This unarmed test aircraft is also fitted with the 'Spanner I' infra-red search device in the fuselage nose. The 'Spanner' device, experimentally used during the early stages of night fighting, did not prove itself on operations.*

Top: *For the acquisition of aerial targets some night fighting areas of the 'Kammhuber Line' had the FuMG 65 'Würzburg-Riese' (Giant Würzburg) ground radars which had a range of 60km (37 miles). They were mounted on a concrete base and rotated through 360°. The external metal ladder leads to the box-shaped operations room. The 'Giant Würzburg' was the forerunner of the later radio telescopes at Jodrell Bank and elsewhere, as clearly shown by its construction.*

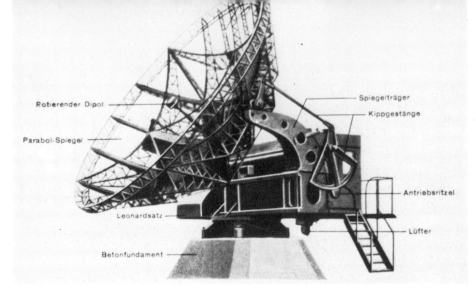

Centre: *As 'Father of Luftwaffe nightfighting'* Gen Ltn *Josef Kammhuber moved into a new technical field in 1941, but for a time demonstrated an ability to adapt German defences to enemy night penetration tactics.*

Bottom: *Two famous names in German night fighting.* Major *Werner Streib (left), commander of I/NJG 1 deployed in the West, greets* Hptm *Prinz zu Sayn-Wittgenstein, commander of I/NJG 100 operational on the Eastern Front, in spring 1943.* [Werner Streib *went on to achieve 65 confirmed night victories and survived the war. Prince Heinrich zu Sayn-Wittgenstein achieved 83 confirmed night victories before being vanquished by an RAF long-range night fighter on the night of 21 January 1944, just after he had shot down his fifth four-engined bomber that night. Tr.*]

Top: *Under the command of Obstl Harlinghousen, 'Fliegerführer Atlantik', the four-engined FW 200s kept watch on and attacked Allied shipping in the Eastern Atlantic. Based at Bordeaux-Merignac, the FW 200C–3s of I/KG 40 covered the main Allied shipping lines. Described by Churchill as the 'Scourge of the Atlantic' in 1941, the FW 200C Condor, a converted commercial airliner, was quite successful for about a year. They generally operated in a wide arc, taking off from France and then landing in Norway, or vice-versa.*

Centre: *The black camouflage of this Ju 88A–4 of Kü.Fl.Gr.106 characterizes the nature of attacks against England in spring 1942. The reinforced British fighter defences had forced the Luftwaffe bombers to fly their penetration raids by night.*

Bottom: *The attempted British landing at Dieppe on 19 August 1942 was frustrated with the active participation of the Luftwaffe. After the battle, the stretch of coast at Dieppe was covered with wrecked landing craft and vehicles.*

Top: *Carrying a 300-litre drop tank and fitted with an automatic film camera, the Bf 109E–5 was used for 'forced reconnaissance' over England. The 'yellow F' belongs to 3.(F)/123 based at Brest-South in July 1942.*

Centre: *A jubilee at 1.(F)/123, the so-called 'Drunken Head squadron'. The crew who has just completed the squadron's 500th operational flight over England is back again at Buc and being congratulated by their commander. From left: Oblt Berndt (observer), Oblt Rüeck (pilot), Ofw Reiche (radio operator) and Uffz Reiser (gunner).*

Bottom: *The engines of a Do 217E–2 of II/KG 40 are being run up at Soesterberg in Holland. Although as evidenced from the camouflage of this aircraft, the Gruppe was operational under 'Fliegerführer Atlantic', it also participated in raids on England in November 1941.*

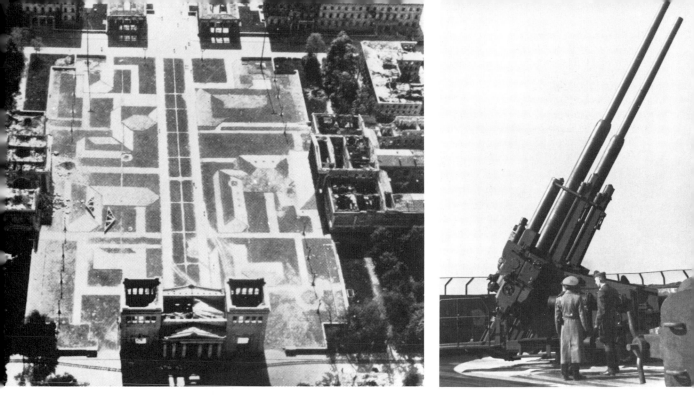

Top left: *To deceive Allied reconnaissance aircraft prominent open squares in many German towns were covered by camouflage netting or disguised by paintwork. In 1944 the Königsplatz in Munich was painted over to look like house roofs.*

Top right: *The increasing number of bombing raids by the RAF and US Eighth Air Force demanded more effort from the anti-aircraft defences in the German homeland. In some instances anti-aircraft gun batteries were fitted on special 'Flakbunkers' situated in the centre of larger towns, like this twin-barrel 128mm heavy Flak gun atop the Zoo Flakbunker in Berlin.*

Bottom: *Passive defences against low-level attacks were another matter. These were mainly used around port installations; an example was this barrage balloon detachment of the German Navy at Cherbourg in March 1942. The balloons in the hangar are of the $77m^3$ (272cu. ft) type.*

Top: *The large closed formations of American B–17 and B–24 bombers were countered by the Luftwaffe with specially armed aircraft. Here is a Bf 109G–6 fitted with Rüstsatz 2 (Armament set 2), two launching tubes for 210mm Army-type rocket mortar shells. The explosive force of these rocket shells was intended mainly to break up tight bomber formations.*

Centre: *After the introduction of the FW 190 most of the fighter formations in the West re-equipped with this type, as did JG 1. This FW 190A–4 flown by the Technical Officer of II/JG 1, Oblt Hans Mohr, in March 1943 at Woendrecht. Mohr was killed in air combat only a month later, on 16 April 1943.*

Bottom: *Faces are still marked by the strain of aerial combat as one participant describes the loss of Ofw Eberhardt on 2 May 1943. From left:* Lt Tschira, Uffz Ritter and Ofw Hutter – all from 5./JG 1.

Top: *A Ju 88C–6 of 5./NJG 2 over Holland. This night fighter is equipped with the FuG 212 Lichtenstein C–1 air interception radar in the nose and fitted with flame dampers over the engine exhaust stubs.*

Centre: *Two experienced night fighter pilots in a serious mood at Deelen in Holland.* Major *Hans-Joachim Jabs (centre) eventually scored 50 confirmed victories (31 by night), while* Oblt *Heinz-Wolfgang Schnaufer (left) achieved the record figure of 121 confirmed night victories, ending the war as the highest-scoring night fighter pilot ever. [Note the partly covered fuselage of an American B–17 'Flying Fortress' bomber in the background. This could explain the serious looks: once the US day bombers had started penetrating German air space the night fighters too were ordered to attack them. Used to completely different tactics at night, the Luftwaffe night fighter pilots suffered heavily in daytime. The shot-down US bomber may well have served as a display object to examine its defences and possible 'weak' spots. Tr.]*

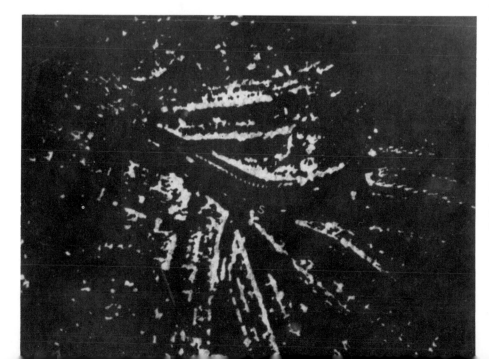

Bottom: *The radar picture of Hamburg, as seen by RAF 'pathfinders' on the night of 24/25 July 1943, which marked the beginning of 'Operation Gomorrha', the fire storm raids on this large German town. The various wet docks stand out clearly as black spaces against the lighter reflections of buildings.*

213

Top: *To hold their own against the four-engined Allied bombers, the calibres and number of weapons carried by German fighters had to be increased. The depicted Bf 110G–2 'Pulkzerstörer' (formation destroyer) is armed with 1 × 37mm BK 3, 7 cannon under the fuselage and 2 × 20mm MG 151/20 cannon in the nose.*

Centre: *Another way of increasing the fire power of the Bf 110 was to fit it with four launching tubes for the 210mm Army-type rocket mortar shells. During the daylight raid on Schweinfurt on 14 October 1943 the American Eighth Air Force lost a considerable numbers of bombers blown out of formation by these high-explosive rocket shells.*

Bottom: *The inside of the radio-direction-finding vehicle at Leeuwarden airfield in Holland. With the help of the revolving direction-finding loop (right) it was possible to keep constant track of friendly aircraft on an operational flight.*

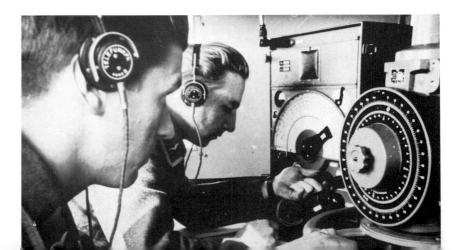

Top: *From summer 1943 onwards the Allied daylight bombing raids began to concentrate more on aircraft production plants and airfields. On 17 August 1943 a 'bomb carpet' ploughed up the airfield and installations at Istres, near Marseille.*

Centre: *The Focke-Wulf assembly plant near Marienburg " in East Prussia was hit by an American daylight precision bombing raid on 10 October 1943 which knocked it out of action for several weeks. Such accurate concentrated bombing raids necessitated dispersal of aircraft production facilities, some of them in underground works.*

Bottom: *The invasion in the West is now expected daily. This picture was taken after one of the frequent discussions of the current situation among the commanders of individual fighter formations in May 1944. From left: Obstl Gustav Roedel (JG 27), Hptm Fritz Kath, Major Heinz Bär (JG 3) and Major Kurt Bühligen (JG 2).*

215

Top: *The crews of 'Pulkzerstörer' (formation destroyers) were advised to wear steel helmets over their ordinary flying helmets as additional protection against the massed defensive firepower of the American bomber formations. Ofw Fritz Buchholz of 6./ZG 26 in his Me 410 3U+EP at Königsberg/ Neumark in May 1944.*

Centre: Lt *Peter Bauer of 3./ZG 76 in a Me 410B–2/U2/R5. The standard armament of this 'Zerstörer' version consisted of eight 20mm MG 151/20 cannon and two 7.92mm MG 17 machine guns. Under ideal conditions, this concentrated firepower could literally 'saw' a four-engined bomber to bits.*

Bottom: *The lasting effects of area bombing are clearly shown in this view of destroyed accommodation buildings at the Quackenbrück airfield after a raid by American four-engined bombers on 8 April 1944. Note the up-ended staff car in the centre background.*

Top: *When finally produced in series, the He 177A Greif heavy bomber became available for attacks on England in 1944. After initial structural defects had been seen to, the He 177 became a useful and reliable combat aircraft. Shown here are two He 177A–5 Greif bombers at München-Riem in March 1944.* [*Due to its early predilection for catching fire in the air the He 177 was derisively named the 'Luftwaffe firelighter'. However, once the two coupled 24-cylinder engine installations had been redesigned and the fuel and oil leads better protected, the fire risk was banished.* Tr.]

Centre: 'Angriffsführer England', *the 'Attack leader England'*, Obstl *Dietrich Pelz. He was entrusted by the Luftwaffe Commander-in-Chief, Göring, with the coordination of 'Operation Steinbock'. Known in the UK as the 'Baby Blitz', it envisaged a series of revenge attacks on British towns beginning January 1944.*

Bottom: *A 'field modification' of the He 177A tail-gun position. The normal tail armament of one 20mm MG 151/20 cannon has been replaced by two 13mm MG 131 heavy machine guns for greater volume of fire.*

217

Top: *Compared to its predecessor, the Do 217E, the crew accommodation of the new Do 217M shows a more attractive aerodynamic shape. This version was delivered to some bomber Gruppen beginning autumn 1942.*

Centre: *The 'illuminator' Gruppe, I/KG 66 was equipped with the Ju 88S, a more streamlined and faster version of the basic design. During 'Operation Steinbock' (the 'Baby Blitz') this unit would fly ahead of the bomber formations and mark the targets. The J88S–1 Z6+LH of 1./KG 66 at Montdidier late in September 1944.*

Bottom: *An impressive night photograph of a Ju 188E of 10./KG 2 taxiing to its take-off position at Villaroche. This particular bomber carries four AB 250 bomb containers under its wing roots. Each AB 250 contained 17 × SD 10A (10kg/22lb) splinter bombs and weighed 215kg (474lb). The AB 250 containers were intended for use mainly against personnel, motor convoys, parked aircraft and similar targets.*

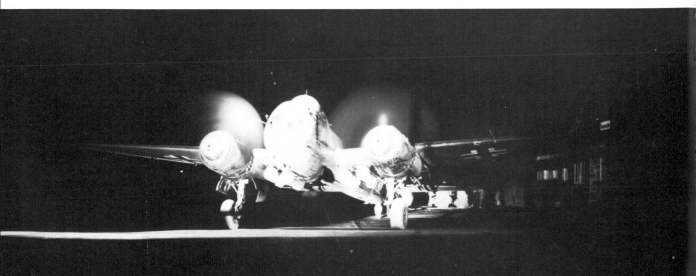

Top: *The long awaited Ju 88G night fighters have finally arrived at II/NJG 2. Faster and with extended flight endurance, this new aircraft allowed the crews to 'swim' longer with the bomber stream and combat more radar contacts. A Ju 88G–1 at Kassel-Rothwesten.*

Centre: *The highest victory series in one night was achieved by the crew of Oblt 'Tino' Becker of I/NJG 6: on the night of 14/15 March 1945 they shot down nine four-engined RAF bombers. From left:* Hptm *Martin Becker (pilot),* Objetr *Weizenbach (gunner) and* Ofw *Karl L. Johanssen (radar operator).*

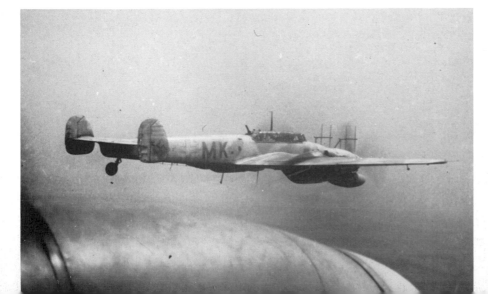

Bottom: *The Bf 110G–6 piloted by* Oblt *Becker during a twilight flight in summer 1944. The aircraft is equipped with the FuG 220 (Lichtenstein SN–2) air interception radar and carries additional under-wing fuel tanks. The two small protuberances on the rear cockpit cover hide the gun muzzles of the 'Schräge Musik' ('Slanted music'), two obliquely-fitted 20mm MG FF cannon.* Hptm *Martin Becker (as he became) ended the war with 58 confirmed night victories.*

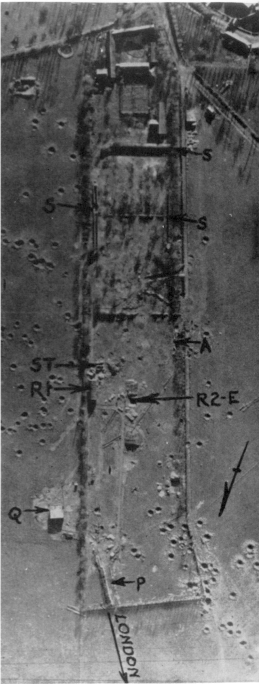

Top left: *A newly delivered V1 'Kirschkern' (Cherry stone – cover name for the V1) is being driven into the storage bunker. Flying at a speed of 650km/h (404mph), the V1 was relatively slow and many did not reach their targets, falling victim to the anti-aircraft barrages and intercepting fighters on the way. [The British AA guns were the first to use the VT or proximity fuse, manufactured in the USA and delivered to Britain specially to combat the V1s – then a great threat. These VT fuses were later also instrumental in defeating the 'Kamikaze' pilots, but their use was kept secret until after the end of the war – when the bigger news about the A-bomb completely overshadowed the news releases about the important role played by the proximity fuse. Tr.]*

Bottom: *Designed as a target defence interceptor, the malicious rocket-powered Me 163 Komet had only a limited flight endurance and equipped only one Luftwaffe formation, JG 400. From summer 1944 onwards, its squadrons were deployed to protect the hydrogenation plants in the Merseburg-Leipzig area.*

Top: *Identified by Allied aerial reconnaissance, the V1 launching installations in the area between Boulogne and Calais were under constant surveillance and bombing attacks began in spring 1944. An aerial reconnaisance photograph of the V1 firing installation at St Josse-en-Bois after an Allied bombing raid on 28 February 1944.*

Top left: *The task of ferrying new aircraft from factories to the fighting troops was carried out by Flugzeugüberführungsgeschwader 1 (1 Aircraft Ferrying Geschwader) commanded by Obstl Zeidler. Among the pilots of this formation were also a number of famous German airwomen volunteers. One of them was the well-known pre-war sports pilot Vera von Bissig, who served with the Gruppe Centre.*

Top right: *Training another woman ferry pilot on the Bf 109 'Gustav': Ofw Setzpfand (in the instructor's seat) acquaints Beate Uhse (front seat) with the characteristics of this type in a two-seat Bf 109G–12 at Jüterbog in August 1944.*

Bottom: *A line-up of Bf 109G–14/AS fighters newly ferried by Fl.Ü.G.1 to III/JG 1 'Oesau' at Anklam. At this time, December 1944, many newly completed aircraft were no longer delivered in the usual camouflage colours but sprayed overall in standard blue-grey (Luftwaffe colour No. 76).*

Top: *Among the tasks of Versuchsverband des Ob.d.L. (Luftwaffe High Command Experimental Detachment) were comparison demonstrations of captured enemy aircraft to various fighter Gruppen. In this way, the Luftwaffe fighter pilots had an opportunity to get to know the manoeuvrabilty of their opponents in practice. The P–51C Mustang T9+CK at Paderborn in autumn 1944.*

Right: *Nearly all force-landed but still-serviceable four-engined Allied bombers went to KG 200, a special Luftwaffe formation. Before that, the Luftwaffe fighter pilot trainees at the Villacoublay fighter-training school would carry out practice attacks on this Boeing B–17E Flying Fortress with the Luftwaffe registration DL+XC, photographed in July 1943. This B–17E, US Serial 124585, was formerly 'Wolf Hound' of 360th BS/303rd Bomb Group, US Eighth AF, forced down near Rouen in France on 12 December 1942 – the first B–17 to fall almost intact into German hands.*

Bottom: *A B–24J Liberator of KG 200 snapped during a stop-over landing at Bergamo in Italy. Completely resprayed and carrying full Luftwaffe markings, these long-range aircraft were especially suitable for nocturnal agent-dropping flights far into the enemy rear areas.*

Top: *With the 'long-nose Dora', the FW 190D-9 powered by a Jumo 213 inline engine, some* Luftwaffe fighter *Gruppen received an aircraft in summer 1944 that was equal in high-altitude performance to the American P-51D Mustang.*

Left: *Great hopes and expectations rested on the Me 262, the world's first operational jet fighter. However, due to delays in the development phase and consequently late introduction into service – but above all because of the overwhelming Allied aerial superiority over Germany, this turbo-jet fighter could no longer reverse the tide of events.*

Bottom: *As before, the 'old warriors' of the day fighter arm had to carry the main burden in Germany's defence. Lt Walter Brandt, the* Staffelkapitän *of 2./JG 3 'Udet', who continued flying with an amputated leg, relaxes between two operational sorties playing cards with his wingman, Fw Rudolf Hener at Erfurt-Bindersleben in autumn 1944.*

Top: *Improved camouflage measures on airfields after the Allied invasion of 6 June 1944: with an added 'step' in the roof and mock lateral outbuildings this hangar at Schiphol becomes a field barn.*

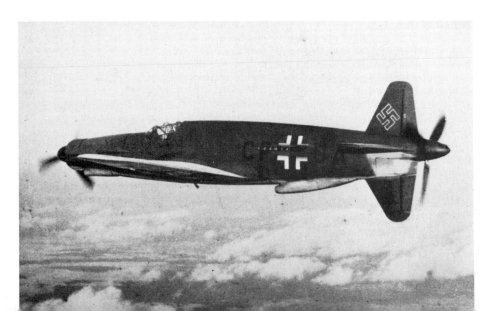

Right: *On 10 November 1944 an American 'bomb carpet' was laid on the Luftwaffe airfield at Wiesbaden-Erbenheim. It required weeks of clearing and repair work before the flying operations from this airfield could be back to normal.*

Bottom: *The Do 335 Pfeil (Arrow), powered by tandem engines. [The Do 335 had a longish development period, and problems with cooling of the rear engine; it was also a rather big and expensive aircraft for a single-seat fighter. It was first flown in September 1943; the first Do 335A–O pre-production aircraft was delivered for operational trials in July 1944, and series production began in October 1944. The first Do 335A–1 production aircraft were ready in March 1945. Altogether 37 Do 335s were built, including 20 prototypes. A special Ekdo 335 (335 Test Detachment) was formed in spring 1945 for operational trials, and some Do 335s were delivered – if not used in action. Shown here is the Do 335V–1 CP+UA. Tr.]*

Top: *The night fighters waited in vain! The twin-engined He 219A–5 Uhu (Owl) which equipped only one unit, II/NJG 1. Difficulties during the initial stages of its production moved the office of aircraft procurement to delete the He 219 from the programme.*

Centre: *The Tank Ta 154 bad-weather fighter was intended to become the Luftwaffe counterpart to the British Mosquito, but only about ten examples of the all-wood Ta 154A–1 were delivered in 1945.*

Bottom: *Lacking any newly-developed aircraft, the night fighter crews continued to fly their Ju 88s and Bf 110s. Four pilots of II/NJG 2 at Köln-Butzweilerhof with the wife of their* Staffelkapitän *Brinkhaus. From left: Lt Erich Jung, Frau Brinkhaus, Lt S.Elsässer, Lt L.Jarsch and Lt M. Tschirsch.*

225

Top: *After withdrawing the aircraft park of Fliegerführer Atlantik to Germany, the BV 222C–09 'Wiking' continued serving as transport for several months before she fell victim to a low-level attack by Allied fighters in the port of Travemünde.*

Centre: *Beginning November 1943, the two Staffeln of FAGr.5 with their Ju 290As flew long-range reconnaissance sorties from Mont de Marsant in France to 30° West over the Atlantic. Operating as 'contact aircraft' ('shadowers') to Allied convoys, the Ju 290A crews often remained airborne for 18 hours at a stretch. The Ju290A–3, 9V+DH, of 1./FAGr.5.*

Bottom: *Delivered to an operational unit but not flown in combat. The He 162 'Volksjäger' was designed and put into series production in the record time of just five months. Aircraft of I/JG 1 'Oesau' lined up at Leck ın Holstein in May 1945.*

Top: *The V2 long-range rocket (official designation A4) was an Army development. With its one-ton warhead, it was only one of the so-called 'V-Waffen' (Revenge weapons) that gave the Allies their last fright. An experimental V2 example is shown here on its special transport vehicle, the so-called 'Meillerwagen'.*

Centre: *The 'Father and Son' aircraft too became a talking point during the last months of hostilities. A manned fighter coupled to the lower unmanned component controlled the combination in flight towards the target and released itself after switching on the automatic flight controls of the powered 'flying bomb'. Shown here is a 'Mistel 1' (Mistletoe 1, covername for this novel weapon) combination, a Ju 88 'bomb' with its Bf 109G control aircraft on its cradle at Burg near Magdeburg in April 1945.*

Bottom: *Attaching the 3.8-ton (8377lb) warhead to the airframe of the specially prepared Ju 88. These 'Mistel' combinations were first used against the Allied supply ports in Normandy in June, 1944, and then against the Neisse and Oder bridges in the East and the Rhein bridges in the West during the final months of the war.*

Top: *The Ar 234B Blitz jet bomber and reconnaissance aircraft proved the high technical level of German aerial armament. Aircraft of III/KG 76 were still flying attacks on British supply columns at Bremervörde early in May 1945.*

Centre: *Only by a constant game of 'hide and seek', involving camouflage and the so-called 'shadow' fields, could the Luftwaffe formations protect their aircraft from continuous attacks by Allied fighter-bombers and keep them operational. A Ju 188F of 4.(F)/122 well covered by camouflage netting at Bozen in May 1945.*

Bottom: *Improvisation was the need of the moment. Armed with Panzerfaust rockets the Bü 181 Bestmann trainers were formed into 'tank hunting' squadrons to attack the advancing enemy armour. In most cases these slow aircraft were shot down by ground defences during their approach flight.* [*The* Panzerfaust *was a one-shot German infantry anti-tank rocket, introduced on the Eastern Front in 1943. The Bü 181 is shown carrying four late-model* Panzerfaust *100 hollow-charge rocket shells which could penetrate 200mm of armour at 30° impact at 100m range. Tr.*]

228

Top: *The end of an era. As here at Holzkirchen, the very latest aircraft types were piled up on top of each other in the various collecting centres, awaiting the cutting torch. The Bf 109G on top of the brand-new Me 262 jet fighter carries Hungarian markings. The last Hungarian fighter squadrons fought to the very end, retreating into Austria.*

Centre: *A Soviet infantryman poses proudly for a souvenir photograph of the Great Patriotic War (the correct Russian term) on the wreck of a He 111H bomber at Königgratz.*

Bottom: *The latest German technical developments, such as the Do 335A–12 two-seat trainer version of the Pfeil fighter, became part of a most welcome war booty to the victorious powers, to be taken back home for a thorough technical examination. Other prizes included the very advanced jet- and rocket-propelled aircraft and prototypes, as well as Swept wing, supersonic and other research data.*

Luftwaffe formations: an explanatory note by the translator

As there were no exact equivalents in the wartime RAF and USAAF, this English-language edition has maintained the original German unit designations, such as *Staffel*, *Gruppe* and *Geschwader*.

Depending on its role, a *Staffel* could have an establishment of 12–16 aircraft (including reserves), although from 1942 onwards the average strength was seldom more than 9–12 aircraft, including those undergoing maintenance. The basic operational unit was a *Gruppe*, which would comprise three to four *Staffeln*. A *Geschwader* had a nominal strength of three to four *Gruppen*, of which the IV *Gruppe* was generally a conversion/operational training unit. The actual operational strength of all these formations was usually lower than their establishment due to losses, unserviceable aircraft and outstanding material and personnel replacements, especially during the later phases of the war.

There were also many separate fighter, bomber and transport *Gruppen*, some of them of only temporary existence, while coastal and reconnaissance aviation was organized on *Gruppe* basis, with individual *Staffeln* often operating independently and far away from each other and their *Stab* (headquarters).

A number of such flying formations were organized into a *Fliegerdivision* and *Fliegerkorps*, which would be subordinated to a Luftflotte (Air Fleet) according to operational requirements.

As a rule, the formation designations were used in abbreviated form where the *Staffeln* within a *Geschwader* were indicated by Arabic numerals and the *Gruppen* by Roman numerals Thus, 5./KG 26 = 5.*Staffel* of *Kampfgeschwader* 26 (part of II.*Gruppe*). and I/JG 27 = I.*Gruppe* of *Jagdgeschwader* 27. Separate *Gruppen* and their *Staffeln* were indicated by Arabic numerals, e.g. 3./KGr.806 = 3.*Staffel* of *Kampfgruppe* 806, or 2./Kü.Fl.Gr.106 = 2.*Staffel* of *Küstenfliegergruppe* 106 (coastal aviation).

In addition, some Luftwaffe formations were also named and others known under their descriptive names:

Fighters: JG 1 'Oesau' (after 1943)
JG 2 'Richthofen' (from 1938)
JG 3 'Udet' (from 1942)
JG 5 ('*Eismeergeschwader*')
JG 26 'Schlageter' (from 1939)
JG 27 ('*Afrika-Geschwader*')
JG 51 'Mölders' (from 1942)
JG 53 ('*Pik-As Geschwader*')
JG 54 '*Grünherz-Geschwader*'

Bombers: KG 1 'Hindenburg'
KG 2 ('*Holzhammer-Geschwader*')
KG 3 ('*Blitz-Geschwader*')
KG 4 'General Wever'
KG 26 ('*Löwen-Geschwader*')
KG 27 'Boelcke'
KG 30 ('*Adler-Geschwader*')
KG 51 ('*Edelweiss—Geschwader*')
KG 53 'Legion Condor'
KG 54 ('*Totenkopf-Geschwader*')
KG 55 ('*Greifen-Geschwader*')

Other named flying formations:
Aufkl.Gruppe 10 'Tannenberg'
Stuka-Geschwader 2 'Immelmann'
Zerstörer-Geschwader 1 (ZG 1) ('*Wespen-Geschwader*)
Zerstörer-Geschwader 26 'Horst Wessel'

German abbreviations used in text

Aufkl.Gr.   = *Aufklärungsgruppe* = Reconnaissance *Gruppe*
E.Gr.   = *Erprobungsgruppe* = (Operational) trials *Gruppe*
(F)   = (*Fernaufklärungs-*) = Long-range reconnaissance
FAGr.   = *Fernaufklärungsgruppe* = Long-range reconnaissance *Gruppe*
Fhr   = *Fähnrich* = Officer candidate
Fw   = *Feldwebel* = Sergeant
Gefr   = *Gefreiter* = Lance corporal
Gen.d.Flg   = *General der Flieger*
Gen.Lt   = *Generalleutnant*
Gen.Maj   = *Generalmajor*
Gen.Oberst   = *Generaloberst*
(H)   = (*Heeres-*) = (Army); short-range reconnaissance
Hptm   = *Hauptmann*
JG   = *Jagdgeschwader*, Fighter *Geschwader*
JGr   = *Jagdgruppe* = Fighter *Gruppe*
KG   = *Kampfgeschwader* = Bomber *Geschwader*
KGr   = *Kampfgruppe* = Bomber *Gruppe*
Kü.Fl.Gr.   = *Küstenfliegergruppe* = Coastal aviation *Gruppe*
LG   = *Lehrgeschwader* = Instructional/demonstration *Geschwader*
Lt   = *Leutnant*
Maj   = *Major*
NAGr.   = *Nahaufklärungsgruppe* = Short-range reconnaissance *Gruppe*
NJG   = *Nachtjagdgeschwader* = Night-fighter *Geschwader*
NSGr.   = *Nachtschlachtgruppe* = Night assault *Gruppe*
Ob.d.L.   = *Oberbefehlshaber der Luftwaffe*
Oblt   = *Oberleutnant*
Obstl   = *Oberstleutnant*
Ofw   = *Oberfeldwebel*
SG   = *Schlachtgeschwader* = Close support/ground attack *Geschwader*
SKG   = *Schnellkampfgeschwader* = High-speed bomber *Geschwader*
St.G   = *Sturzkampfgeschwader* = Stuka *Geschwader*
TGr.   = *Transportgruppe*
Uffz   = *Unteroffizier*
ZG   = *Zerstörergeschwader* = Heavy fighter ('Destroyer') *Geschwader*

Comparative airforce ranks

LUFTWAFFE	RAF	USAAF	SOVIET VVS
Reichsmarschall	—	—	—
Generalfeldmarschall	Marshall of the RAF	—	*Glavny Marshal*
Generaloberst	Air Chief Marshal	General (4-star)	*Marshal*
General der Flieger	Air Marshal	Lt Gen (3-star)	*General-Polkovnik*
Generalleutnant	Air Vice Marshal	Maj Gen (2-star)	*General-Leitenant*
Generalmajor	Air Commodore	Brig Gen (1-star)	*General-Mayor*
Oberst	Group Captain	Colonel	*Polkovnik*
Oberstleutnant	Wing Commander	Lt Col	*Podpolkovnik*
Major	Squadron Leader	Major	*Mayor*
Hauptmann	Flight Lieutenant	Captain	*Kapitan*
Oberleutnant	Flying Officer	First Lt	*Starshy Leitenant*
Leutnant	Pilot Officer	Second Lt	*Leitenant*
—	—	Flight Officer	*Mladshy Leitenant*
Stabsfeldwebel	Warrant Officer	—	—
Hauptfeldwebel	—	—	*Starshina*
Oberfeldwebel	Flight Sergeant	Master Sgt (1st grade)	*Starshy Serzhant*
Feldwebel	—	Techn Sgt (2nd grade)	*Serzhant*
Unterfeldwebel	—	Staff Sgt (3rd grade)	—
Unteroffizier	Sergeant	Sergeant (4th grade)	*Mladshy Serzhant*
Stabsgefreiter	—	—	—
Hauptgefreiter	—	—	—
Obergefreiter	Corporal	Corporal (5th grade)	—
Gefreiter	Leading Aircraftman	Private Ist class	*Yefreitor*
Flieger	Aircraftman	Private	*Soldat*

In addition, there were the Luftwaffe ranks of *Fähnrich* and *Oberfähnrich* (Officer Candidate and Senior Officer Candidate), which had no equivalent in the RAF or USAAF.

The Luftwaffe also differed from British and American practice in that a German pilot could be an ordinary *Flieger* or *Gefreiter* whereas an RAF pilot had to be at least a Sergeant, and a USAAF pilot a Second Lieutenant.

A *Staffelkapitän* (*Staffel* leader) was an appointment with command and administrative powers over all other members of that unit, including officers of equal or even more senior rank.